VOLUME NINETY SEVEN

Advances in
PROTEIN CHEMISTRY AND STRUCTURAL BIOLOGY

Metal-Containing Enzymes

VOLUME NINETY SEVEN

ADVANCES IN
PROTEIN CHEMISTRY AND STRUCTURAL BIOLOGY
Metal-Containing Enzymes

Edited by

CHRISTO Z. CHRISTOV
Department of Applied Sciences
Northumbria University at Newcastle
Newcastle-upon-Tyne, United Kingdom

AMSTERDAM • BOSTON • HEIDELBERG • LONDON
NEW YORK • OXFORD • PARIS • SAN DIEGO
SAN FRANCISCO • SINGAPORE • SYDNEY • TOKYO
Academic Press is an imprint of Elsevier

Academic Press is an imprint of Elsevier
225 Wyman Street, Waltham, MA 02451, USA
525 B Street, Suite 1800, San Diego, CA 92101-4495, USA
32 Jamestown Road, London NW1 7BY, UK
The Boulevard, Langford Lane, Kidlington, Oxford OX5 1GB, UK

First edition 2014

Copyright © 2014 Elsevier Inc. All rights reserved.

No part of this publication may be reproduced or transmitted in any form or by any means, electronic or mechanical, including photocopying, recording, or any information storage and retrieval system, without permission in writing from the publisher. Details on how to seek permission, further information about the Publisher's permissions policies and our arrangements with organizations such as the Copyright Clearance Center and the Copyright Licensing Agency, can be found at our website: www.elsevier.com/permissions.

This book and the individual contributions contained in it are protected under copyright by the Publisher (other than as may be noted herein).

Notices
Knowledge and best practice in this field are constantly changing. As new research and experience broaden our understanding, changes in research methods, professional practices, or medical treatment may become necessary.

Practitioners and researchers must always rely on their own experience and knowledge in evaluating and using any information, methods, compounds, or experiments described herein. In using such information or methods they should be mindful of their own safety and the safety of others, including parties for whom they have a professional responsibility.

To the fullest extent of the law, neither the Publisher nor the authors, contributors, or editors, assume any liability for any injury and/or damage to persons or property as a matter of products liability, negligence or otherwise, or from any use or operation of any methods, products, instructions, or ideas contained in the material herein.

ISBN: 978-0-12-800012-0
ISSN: 1876-1623

For information on all Academic Press publications
visit our website at store.elsevier.com

CONTENTS

Contributors   vii
Preface   ix

1. Type-3 Copper Proteins: Recent Advances on Polyphenol Oxidases   1
Cornelia Kaintz, Stephan Gerhard Mauracher, and Annette Rompel

1. Introduction   2
2. Primary Structure and Molecular Weights of PPOs   7
3. Oxo Complex   17
4. X-ray Crystallographic Structural Data of PPOs and Hemocyanins   22
5. Conclusions and Outlook   30

Acknowledgments   30
References   30

2. Biophysical Studies of Matrix Metalloproteinase/Triple-Helix Complexes   37
Gregg B. Fields

1. MMPs and Collagen Hydrolysis   37
2. Structures of Full-Length, Collagenolytic MMPs in Solution and in the Solid State   38
3. Structural Evaluation of MMP Interactions with Collagen   41
4. Mechanism of Collagenolysis   43
5. Heterogeneity in MMP Structures   46

Acknowledgment   47
References   47

3. Catalytic Mechanisms of Metallohydrolases Containing Two Metal Ions   49
Nataša Mitić, Manfredi Miraula, Christopher Selleck, Kieran S. Hadler, Elena Uribe, Marcelo M. Pedroso, and Gerhard Schenk

1. Introduction   50
2. Metallo-β-Lactamases, Major Culprits in the Emergence of Antibiotic Resistance   50
3. Methionine Aminopeptidase, a Target for Novel Anticancer Drugs   57
4. Glycerophosphodiesterase, a Very Promiscuous Potential Bioremediator   60
5. PAP, an Alternative Target to Treat Osteoporosis   65

6. Conclusions and Outlook: Agmatinase, an Emerging Target for
   Biotechnological Applications ... 69
   Acknowledgments ... 71
   References ... 72

4. **Applications of Quantum Mechanical/Molecular Mechanical Methods to the Chemical Insertion Step of DNA and RNA Polymerization** ... 83

   Lalith Perera, William A. Beard, Lee G. Pedersen, and Samuel H. Wilson

   1. Introduction ... 85
   2. Methods for Describing Reactive Pathways ... 85
   3. DNA Polymerase β ... 87
   4. Application of QM/MM to Systems Similar to Pol β ... 98
   5. RNA Polymerase ... 103
   6. Thoughts for Future QM/MM Simulations on Ternary Substrate Complexes of Nucleic Acid Polymerases ... 106
   7. Conclusions ... 109
   Acknowledgments ... 109
   References ... 109

5. **Monitoring the Biomolecular Interactions and the Activity of Zn-Containing Enzymes Involved in Conformational Diseases: Experimental Methods for Therapeutic Purposes** ... 115

   Giuseppe Grasso

   1. Zn-Metalloproteases and Conformational Diseases ... 116
   2. Analytical Techniques Used to Study ZnMPs-Substrate/Inhibitors Interactions ... 120
   3. Conclusions and Future Perspectives ... 135
   References ... 135

Author Index ... *143*
Subject Index ... *159*

CONTRIBUTORS

William A. Beard
Laboratory of Structural Biology, National Institution of Environmental Health Sciences, Research Triangle Park, North Carolina, USA

Gregg B. Fields
Torrey Pines Institute for Molecular Studies, Port St. Lucie, Florida, USA

Giuseppe Grasso
Dipartimento di Scienze Chimiche, Università degli Studi di Catania, Catania, Italy

Kieran S. Hadler
School of Chemistry and Molecular Biosciences, The University of Queensland, Brisbane, Queensland, Australia

Cornelia Kaintz
Institut für Biophysikalische Chemie, Fakultät für Chemie, Universität Wien, 1090-Wien, Austria

Stephan Gerhard Mauracher
Institut für Biophysikalische Chemie, Fakultät für Chemie, Universität Wien, 1090-Wien, Austria

Manfredi Miraula
Department of Chemistry, National University of Ireland, Maynooth, Maynooth, Co. Kildare, Ireland, and School of Chemistry and Molecular Biosciences, The University of Queensland, Brisbane, Queensland, Australia

Nataša Mitić
Department of Chemistry, National University of Ireland, Maynooth, Maynooth, Co. Kildare, Ireland

Lee G. Pedersen
Department of Chemistry, University of North Carolina at Chapel Hill, Chapel Hill, North Carolina, USA

Marcelo M. Pedroso
School of Chemistry and Molecular Biosciences, The University of Queensland, Brisbane, Queensland, Australia

Lalith Perera
Laboratory of Structural Biology, National Institution of Environmental Health Sciences, Research Triangle Park, North Carolina, USA

Annette Rompel
Institut für Biophysikalische Chemie, Fakultät für Chemie, Universität Wien, 1090-Wien, Austria

Gerhard Schenk
School of Chemistry and Molecular Biosciences, The University of Queensland, Brisbane, Queensland, Australia

Christopher Selleck
School of Chemistry and Molecular Biosciences, The University of Queensland, Brisbane, Queensland, Australia

Elena Uribe
Department of Biochemistry and Molecular Biology, University of Concepción, Concepción, Chile

Samuel H. Wilson
Laboratory of Structural Biology, National Institution of Environmental Health Sciences, Research Triangle Park, North Carolina, USA

PREFACE

Bioinorganic chemistry became an independent scientific discipline before more than 50 years and achieved fast growth over the past 20 years. The purification and crystallization of metal-containing enzymes led to the discovery of the crystal structures of large number enzymes in Protein Data Bank. In addition, the development of the spectroscopic methods and massive spectroscopic instruments, including synchrotrons, refined and complemented the picture of how metals interact with the protein ligands and cofactors, providing further insight about the coordination states, geometries, oxidation states, and covalent character of metal sites in proteins. The fast development of theoretical methods in combination with the growth of the computer facilities made possible the application of the quantum mechanics, molecular dynamics, and combined quantum mechanical and molecular mechanical methods for exploring the electronic structure, chemical bonding, and mechanisms of metal-containing enzymes, this way complementing the results made by experimental methods. It is important, however, the theoretical methods to be validated in respect to the spectroscopic and crystallographic data in order to represent accurately the geometric and electronic structure of the metal sites in proteins.

This issue of *APCSB* is centered on the recent advances in metalloenzymology and represents some of the most advanced achievements in the understanding of the structure, structure–functions relationships, mechanisms, and biomedical roles of the metal-containing enzymes. The volume includes contributions on type-3 copper proteins, matrix metalloproteinase, metallohydrolases with two metal ions, metal-containing enzymes participating in nucleic acid biosynthesis, and Zn-containing enzymes involved in conformational diseases.

Christo Z. Christov
Department of Applied Sciences,
Northumbria University at Newcastle,
Newcastle-upon-Tyne,
United Kingdom

CHAPTER ONE

Type-3 Copper Proteins: Recent Advances on Polyphenol Oxidases

Cornelia Kaintz, Stephan Gerhard Mauracher, Annette Rompel[1]

Institut für Biophysikalische Chemie, Fakultät für Chemie, Universität Wien, 1090-Wien, Austria
[1]Corresponding author: e-mail address: annette.rompel@univie.ac.at

Contents

1. Introduction 2
2. Primary Structure and Molecular Weights of PPOs 7
   2.1 General sequence structure 7
   2.2 Core domain, C-terminal domain, and its proteolytic cleavage site 7
   2.3 Conserved amino acid motifs in the core domain 8
   2.4 Conserved amino acid motifs in the C-terminal domain 9
   2.5 Transit peptide and location of PPOs 9
   2.6 Sequence homologies within PPOs 11
   2.7 *In vitro* activation of PPOs 16
   2.8 *In vivo* activation of PPOs 16
   2.9 Mutants of PPOs 16
3. *Oxo* Complex 17
   3.1 *Oxo* complex investigated by UV/vis absorption spectroscopy 18
   3.2 *Oxo* complex investigated by CD spectroscopy 18
   3.3 *Oxo* complex investigated by resonance Raman spectroscopy 18
   3.4 *Oxo* complex of catechol oxidase 19
   3.5 *Oxo* complex of tyrosinase 19
   3.6 *Oxo* complex of aureusidin synthase 22
   3.7 *Oxo* complex of hemocyanin 22
4. X-ray Crystallographic Structural Data of PPOs and Hemocyanins 22
   4.1 Published structures in the protein data bank 22
   4.2 Structural differences of tyrosinases and catechol oxidases 29
5. Conclusions and Outlook 30
Acknowledgments 30
References 30

Abstract

Recent investigations in the study of plant, fungal, and bacterial type-3 copper proteins are reviewed. Focus is given to three enzymes: catechol oxidases (CO), tyrosinases, and aureusidin synthase. CO were mostly found in plants, however, in 2010 the first fungal CO was published. The first plant-originated tyrosinase was published in 2014, before tyrosinases were only reported in fungi, bacteria, and human. Aureusidin synthase from yellow snapdragon (*Antirrhinum majus*) was first published in 2000, as part of yellow flower coloration pathway. In the last years, many important results on type-3 copper enzymes originated from X-ray crystallographic investigations. In addition, studies on site-directed mutagenesis of amino acids around the active site were performed to identify the regions determining monophenolase and/or diphenolase activity. Although X-ray crystallographic structures of CO and tyrosinases are available, many questions like the response for the activation via proteases, sequence-based or structural-based differences between CO, as well as the physiological roles of many polyphenol oxidases still remain to be addressed.

ABBREVIATIONS

AbPPO tyrosinase from *Agaricus bisporus*
AmAS1 aureusidin synthase from *Antirrhinum majus*
AoCO4 catechol oxidase from *Aspergillus oryzae*
A-TYR active tyrosinase
BmTYR tyrosinase from *Bacillus megaterium*
CO catechol oxidase
CT charge transfer
cTP chloroplast transit peptide
EPR electron paramagnetic resonance
fl full-length
HC hemocyanin
IbCO catechol oxidase from *Ipomoea batatas*
kDa kilo Dalton
L-TYR latent tyrosinase
PHC 2′,4′,6′,3,4-pentahydroxychalcone
PPO polyphenol oxidase
PTM posttranslational modification
PTU phenylthiourea
rc recombinant
ScTYR tyrosinase from *Streptomyces castaneoglobisporus*
SDS sodium dodecyl sulfate
SKE serine–lysine–glutamic acid motif
TYR tyrosinase

1. INTRODUCTION

Metalloproteins containing copper are generally assigned to one of three classes based on their structure and spectroscopic properties. The mononuclear type-1 centers, "blue copper proteins" are mostly involved in electron transfer, type-2 centers, "nonblue copper centers" are typically found in enzymes that activate molecular oxygen, and type-3 centers contain two copper atoms and also activate molecular dioxygen. A number of excellent reviews have been published on copper proteins among them Solomon, Baldwin, and Lowery (1992).

The type-3 copper centers bind dioxygen in a characteristic "side-on" bridging mode (μ-η^2:η^2) (Kitajima, Fujisawa, Morooka, & Toriumi, 1989) resulting in the activation of dioxygen. The *oxy* form of the proteins exhibits characteristic spectral features which include a strong antiferromagnetic coupling between the two Cu^{2+} ions which suppresses the electron paramagnetic resonance (EPR)-signal and exhibits a resonance Raman-active O–O stretching vibration at 750 cm^{-1} (Eickman, Himmelwright, & Solomon, 1979; Freedman, Loehr, & Loehr, 1976; Himmelwright, Eickman, LuBien, Lerch, & Solomon, 1980; Solomon et al., 1992; Solomon, Tuczek, Root, & Brown, 1994). The absorption band of the *oxy* form at 345 nm is due to an $O_2^{2-}\left(\pi_\sigma^*\right) \rightarrow Cu(II)\left(d_{x^2-y^2}\right)$ charge transfer (CT) transition. The absorption band at 580 nm corresponds to the second $O_2^{2-}\left(\pi_v^*\right) \rightarrow Cu(II)\left(d_{x^2-y^2}\right)$ CT transition. The *met* form has either one (*met*) or two (*met2*) water molecules bridging the two Cu^{2+} ions in a triangular position and is described as the resting state (Eickman et al., 1979).

Polyphenol oxidases (PPOs) are an important class of type-3 copper enzymes and are found in plants, fungi, and animals (for excellent reviews about PPOs, see Mayer (2006); Mayer and Harel (1979)). Type-3 copper-based PPOs are classified by enzyme nomenclature into catechol oxidases (CO, *o*-diphenol:oxygen oxidoreductase, EC 1.10.3.1), tyrosinases (monophenol, *o*-diphenol: oxygen oxidoreductase, EC 1.14.18.1 and *o*-diphenol:oxygen oxidoreductase, EC 1.10.3.1) and aureusidin synthase (2′,4,4′,6′-tetrahydroxychalcone 4′-O-β-D-glucoside:oxygen oxidoreductase, EC 1.21.3.6). CO catalyze the reaction of *o*-diphenols to *o*-quinones, called diphenolase or CO activity (Fig. 1). In contrast, tyrosinases are bifunctional PPOs which catalyze two different reaction types: (1) the *ortho*-hydroxylation of monophenols (monophenolase activity) and (2) the oxidation of *o*-diphenols to *o*-quinones (diphenolase activity), both reactions with electron transfer to oxygen (Fig. 1). Aureusidin synthase (AmAS1) is also a bifunctional, chalcone-specific PPO and catalyzes the hydroxylation and/or oxidation of 2′,4′,6′,3,4-pentahydroxychalcone (PHC) to aureusidin and bracteatin, in an approximate molar ratio of 6:1 (Fig. 2) (Nakayama et al., 2000). Recently the cloning and functional expression in *E. coli* of a polyphenol oxidase transcript from *Coreopsis grandiflora* involved in

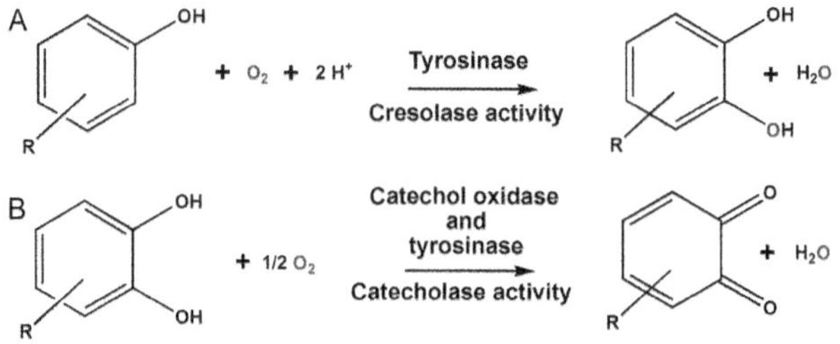

Figure 1 Reactions catalyzed by (A) tyrosinases and (B) tyrosinases and catechol oxidase.

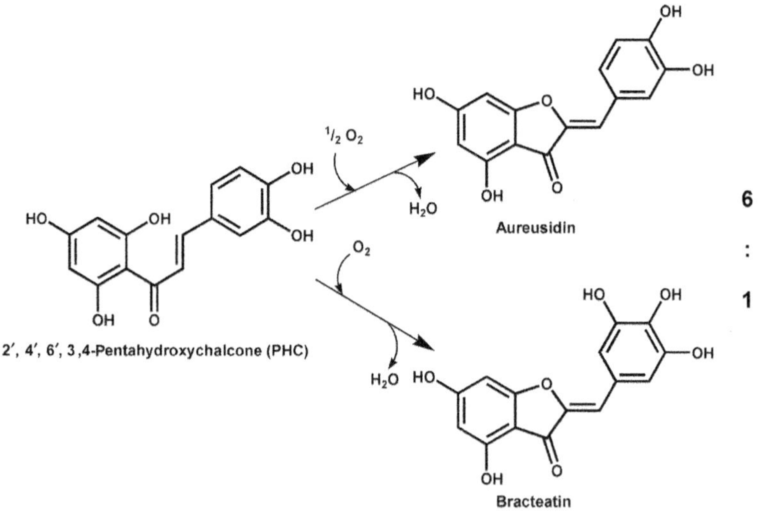

Figure 2 Reaction catalyzed by aureusidin synthase according to Nakayama et al. (2000).

aurone formation was reported (Kaintz et al., 2014). Hemocyanin (HC) serves as an oxygen carrier protein. Decker and Tuczek (2000) demonstrated CO activity, when hemocyanins (both forms of arthropod and mollusc) were subjected to proteolytic treatment detergents, such as sodium dodecyl sulfate (SDS) (*in vitro*). Arthropod hemocyanin is composed of hexameric subunits, each subunit (with a molecular mass of ~75 kilo Dalton (kDa)) can bind one molecule of oxygen. Mollusc hemocyanin consists of one or two 10-mer subunits (with a molecular mass of ~350–450 kDa, folded into seven or eight functional units of ~50 kDa) that assemble into a cylinder-like quaternary structure. Each functional unit can bind one molecule of oxygen (Gaykema et al., 1984; Hazes et al., 1993; Linzen et al., 1985).

COs are well known in plants and recently reported to exist in the sac fungi *Aspergillus oryzae* (Gasparetti et al., 2010), while tyrosinases are identified in humans, plants, and mushrooms, see reviews of Mayer (2006); Mayer and Harel (1979). Aureusidin synthase (AmAS1) is so far only reported in petals of snapdragon (*Antirrhinum majus*) (Nakayama et al., 2000). Many PPOs have a C-terminal domain in the premature state, suggesting that PPOs are expressed as inactive precursors (proenzymes or zymogens) to avoid undesirable intracellular reactions. The active PPO is formed by C-terminal proteolysis of the larger precursor protein (Fujieda, Murata, et al., 2013; Marusek, Trobaugh, Flurkey, & Inlow, 2006; Tran, Taylor, & Constabel, 2012). Several hundreds of articles are published each year on the topic of PPO, CO, and tyrosinase, whereas a countable number of papers are published on the more specialized enzymes such as aureusidin synthase.

PPOs are quite diverse depending on their origin. However, they all have in common highly conserved copper binding regions, with conserved histidine binding motifs around the copper A (CuA) and copper B (CuB) site (Fig. 3). So far, three three-dimensional CO structures are available: two from plants (sweet potato and grape) (Klabunde, Eicken, Sacchettini, & Krebs, 1998; Virador et al., 2010) and one from sac fungi (*A. oryzae*) (Hakulinen, Gasparetti, Kaljunen, Kruus, & Rouvinen, 2013). X-ray crystallographic structures of tyrosinases are available from two

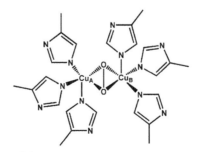

Figure 3 The active site of the *oxy* type-3 copper center.

recombinantly expressed bacterial tyrosinases (Matoba, Kumagai, Yamamoto, Yoshitsu, & Sugiyama, 2006; Sendovski, Kanteev, Ben-Yosef, Adir, & Fishman, 2011), one recombinantly expressed fungal protyrosinase (Fujieda, Yabuta, et al., 2013), the active PPO3 and latent PPO4 fungal tyrosinase from mushrooms (Ismaya et al., 2011; Mauracher, Molitor, Al-Oweini, Kortz, & Rompel, 2014a, 2014b). One insect prophenol oxidase structure (Li, Wang, Jiang, & Deng, 2009) as well as several hemocyanin structures from arthropods and molluscs have been determined (among them Cuff, Miller, van Holde, & Hendrickson, 1998; Jaenicke, Büchler, Decker, Markl, & Schröder, 2011).

Melanins generated by polymerization of quinones are dying compounds, which are responsible for the browning of fruits and vegetables (Martinez & Whitaker, 1995; Yoruk & Marshall, 2003). Intensive research on PPOs is of interest since the impairment and aging-related browning of agricultural products represents a concern in the food industry (Altunkaya & Gökmen, 2008; Coetzer, Corsini, Love, Pavek, & Tumer, 2001; Halaouli, Asther, Sigoillot, Hamdi, & Lomascolo, 2006; Queiroz, da Silva, Lopes, Fialho, & Valente-Mesquita, 2011; Spagna, Barbagallo, Chisari, & Branca, 2005).

Human tyrosinase plays a role in the synthesis of melanins and has been proposed to take part in the oxidative chemistry related to Parkinson disease (Tessari et al., 2008). Tessari et al. (2008) presented the *in vitro* reactivity of tyrosinase with α-synuclein (a protein whose function is poorly understood, although its gene has been associated to familiar forms of Parkinson disease) and a cytotoxic model which includes a possible new toxic role for α-synuclein exacerbated by its direct chemical modification by tyrosinase.

Of all type-3 copper enzymes, the physiological role of aureusidin synthase is best understood as AmAS1 revealed a strict correlation with the aurone formation during flower development in Scrophulariaceae (now Plantaginaceae) (Nakayama et al., 2000; Strack & Schliemann, 2001). The role of this enzyme in the biosynthesis of aurones in Scrophulariaceae was demonstrated (Nakayama, 2002; Nakayama et al., 2000, 2001).

This review attempts to work out sequencial, spectroscopical, and structural similarities and differences of PPOs from various origins, especially of CO, tyrosinases, and aureusidin synthase. Most attention is paid to the literature after 2006, when the first crystal structure of a bacterial tyrosinase became available.

2. PRIMARY STRUCTURE AND MOLECULAR WEIGHTS OF PPOs

2.1. General sequence structure

All plant PPOs consists of a core domain, a C-terminal domain, and an N-terminal transit peptide. The core domain and the C-terminal domain together build the latent enzyme, sometimes also called proenzyme. All plant PPOs are expressed *in vivo* as inactive, latent proteins, before being proteolytically activated by removing the C-terminal domain (Marusek et al., 2006; Tran et al., 2012). The N-terminal transit peptide determines the location of the enzyme, most prominently targeted to the thylakoid or vacuole.

2.2. Core domain, C-terminal domain, and its proteolytic cleavage site

The core domains of plant PPOs range from 1050 to 1200 base pairs encoding for 350 to 400 amino acids, resulting in a molecular mass of approximately 38–44 kDa for the active form of the enzyme. The cleavage site found in several plant PPOs is a conserved SKE (serine–lysine–glutamic acid) motif (Flurkey & Inlow, 2008). A SK motif exists in CO of *Vitis vinifera* at S^{353} (Uniprot P43311) (Virador et al., 2010) and in both CO isoenzymes of *Ipomoea batatas* at S^{347} (40 kDa, Uniprot Q9MB14) (Gerdemann et al., 2001) and S^{343} (39 kDa, Uniprot Q9ZP19) (Eicken, Zippel, Büldt-Karentzopoulos, & Krebs, 1998). A fungal extracellular catechol oxidase from *A. oryzae* (AoCO4) (filamentous fungus growing on rice, Uniprot number Q2UNF9) has been reported in 2010, which lacks the C-terminal domain and exhibits a molecular mass of 44 kDa for the core domain (Gasparetti et al., 2010).

Today six isoforms of tyrosinase (AbPPO) from common mushroom (*Agaricus bisporus*) are known (Weijn, Bastiaan-Net, Wichers, & Mes, 2013; Wichers et al., 2003; Wu et al., 2010). Wichers et al. (2003) cloned the total cDNA sequence of AbPPO1 (Uniprot number Q00024) and AbPPO2 (Uniprot number O42713) with sizes of 1.9 and 1.8 kb, respectively, encoding for latent enzymes with a mass of 64 kDa, while active AbPPO1 and AbPPO2 exhibited a molecular mass of 43 kDa (Wichers, Gerritsen, & Chapelon, 1996). The amino acid sequences of AbPPO3 (Uniprot number C7FF04) and AbPPO4 (Uniprot number C7FF05) lead to molecular masses of 66.3 kDa and 68.3 kDa, respectively, and were reported by Wu et al. (2010). AbPPOs 1–4 are cleaved four amino acids after YG (tyrosine–glycine) leading to an active enzyme with a molecular mass of 41–43 kDa (Wu et al., 2010). So far, the amino acid sequences of AbPPO5 and AbPPO6 (Uniprot numbers not available) were only predicted by gene modeling (Weijn et al., 2013). The highly conserved YG motif (found in AbPPO1–4) is also present in PPO5 and PPO6, in positions G^{392} and G^{422}, respectively.

Mushroom tyrosinase from *Pholiota nameko* (Uniprot number A7BHQ9) was sequenced by cDNA cloning encoding for 625 amino acids, leading to a molecular mass of 68 kDa. The active form was isolated from fruit bodies with a molecular size of 42 kDa, cleaved after SVF^{387} (Kawamura-Konishi et al., 2007). Halaouli et al. (2005) isolated fungal tyrosinase from *Pycnoporus sanguineus* in its active form exhibiting a molecular size of 45 kDa (Uniprot Q2TL94). The latent form of tyrosinase from *P. sanguineus* exhibits a mass of 68 kDa, and the cleavage position is not determined yet. The latent form of fungal tyrosinase from *Neurospora crassa* (Uniprot number P00440) exhibits a mass of 75 kDa, and the active form was isolated with a mass of 46 kDa, cleaved after NVF^{407} (Kupper, Niedermann, Travaglini, & Lerch, 1989). Besides, the protyrosinase from *A. oryzae* (Uniprot number Q00234) was recombinantly expressed as latent enzyme with a molecular mass of 67 kDa (Fujita, Uraga, & Ichisima, 1995). The molecular mass of the active enzyme was not determined in that work.

2.3. Conserved amino acid motifs in the core domain

The core domain of all PPOs consists of several highly conserved amino acids and domains. All plant PPOs start with xP (x is a variable amino acid and P is proline). The Cu-binding domains are characterized by conserved histidine residues, beginning most often in a $H_{CuA}CAYC$ motif (Tran et al., 2012). The second Cys in this motif can form a thioether bond with the second conserved histidine of the CuA domain (Tran et al., 2012). In fungi, the first histidine of the CuA-binding site is found in the $H_{CuA}GxP$ motif (Wu et al., 2010). In both fungal and plant PPOs, the third conserved histidine of the CuA site always follows nine positions after the second conserved one.

In the CuB (Figs. 3 and 5) region for both plants and fungi, the three coordinating histidines and the fourth noncoordinating histidine form the motif $H_{CuB}xxxH_{CuB}x(n)FxxH_{CuB}H$ (Tran et al., 2012; Wu et al., 2010). In PPOs where proteolytic processing of the C-terminus has been documented the cleavage occurs immediately C-terminal to the tyrosine YxY motif (Tran et al., 2012).

2.4. Conserved amino acid motifs in the C-terminal domain

The proteolytically cleaved C-terminal domain consists of a polypeptide of approximately 16–18 kDa and its function is widely unknown (Flurkey & Inlow, 2008; Tran et al., 2012). In the C-terminal domain in plant PPOs, a conserved KFDV (lysine–phenylalanine–aspartic acid–valine) motif is described. A conserved CxxC motif was found by Fujieda, Yabuta, et al. (2013) tyrosinase of *A. oryzae* on loop 3 in the C-terminal domain, and in AbPPO2, AbPPO3, and AbPPO4 (Mauracher, Molitor, Michael, et al., 2014). This motif is also observed in the copper chaperones Atox1 and Ccc2 in yeast (Fujieda, Yabuta, et al., 2013; Robinson & Winge, 2010). While the wild type of the melB protyrosinase was successfully copper metalated, C92A mutants and C92A/C522A/C525A mutants had almost no ability to incorporate the copper ions from the cytoplasm of *Escherichia coli*. Thus, Fujieda, Yabuta, et al. (2013) considered that Cys^{92} as well as the $C^{522}xxC^{525}$ motif contribute to the copper incorporation of melB apo-protyrosinase.

2.5. Transit peptide and location of PPOs

The transit peptide of plant originated CO, tyrosinases, and aureusidin synthase has a variable length ranging from 60 to 110 amino acids. The first 35 residues contain a high proportion of serine residues, typical of the stromal peptide of the cTP (chloroplast transit peptide) (Tran et al., 2012). Tran et al. also reported that a so-called thylakoid transfer domain (TTD), which is rich in hydrophobic residues and resembles signal peptides of secretory proteins (Douwe de Boer & Weisbeek, 1991; Endo, Kawamura, & Nakai, 1992), is often evident as well as an alanine cleavage motif (AxA).

Predictions of the locations of various selected plant PPOs were determined by ChloroP 1.1 (Emanuelsson, Nielsen, & von Heijne, 1999) and TargetP 1.1 (Emanuelsson, Nielsen, Brunak, & von Heijne, 2000) and are shown in Tables 1 and 2. Most plant PPOs contain a cTP suggesting that those are transported to the lumen of the thylakoid membranes in chloroplast (Tran et al., 2012). However, a few plant-originated PPOs such as AmAS1 and CO from *Oryzae sativa* do not contain a cTP. There is no information about the intracellular localization of CO from *O. sativa* published until now. Aureusidin synthase is believed to occur in the vacuole, because

Table 1 TargetP 1.1 prediction of transit peptides

UniProt acc. num.	Length	cTP	mTP	SP	Other	RC
I7HUF2	600	0.814	0.166	0.023	0.094	C
B9VS06	595	0.890	0.063	0.032	0.243	C
Q9FRX6	562	0.048	0.096	0.022	0.946	–
P43311	607	0.928	0.034	0.014	0.085	C
Q9MB14	588	0.924	0.103	0.010	0.080	C
G0ZAL2	590	0.933	0.065	0.005	0.202	C
P43309	593	0.969	0.025	0.017	0.115	C
Q948S3	593	0.972	0.023	0.015	0.119	C
Q08307	587	0.968	0.063	0.007	0.070	C
Q08296	585	0.952	0.106	0.007	0.067	C
Q06355	588	0.933	0.080	0.011	0.100	C
F1DBB9	584	0.858	0.151	0.020	0.052	C
O49912	592	0.928	0.058	0.013	0.131	C
R4JDV1	590	0.954	0.057	0.012	0.101	C
C0LU17	603	0.811	0.084	0.025	0.089	C
Q6UIL3	584	0.782	0.207	0.025	0.047	C
D1KEC6	556	0.151	0.265	0.027	0.414	–
Q6YHK5	577	0.929	0.153	0.017	0.047	C
Q06215	606	0.962	0.014	0.021	0.066	C
P43310	639	0.809	0.066	0.060	0.132	C

cTP is the probability of a chloroplast transit peptide, mTP is the probability of a mitochondrial targeting peptide, SP is the probability of a signal peptide, other means the probability of any other location, RC is the highest reliability class and indicates the strongest prediction of the location.

$4'$-glucosides of $2',4',6',4$-tetrahydroxychalcone (THC) and $2',4',6',3,4$-pentahydroxychalcone (PHC), the substrates for the enzyme, are found in vacuoles, and the enzyme activity shows pH-optimum at the pH typical values for vacuoles as described by Nakayama et al. (2000).

In general, fungal tyrosinases do not contain a transit peptide. They are supposed to be cytoplasmic enzymes; however, some fungal tyrosinases were also targeted in the extracellular matrix or in the mycelium (Mayer, 2006).

Table 2 ChloroP 1.1 prediction of presence of a chloroplast transit peptide

UniProt acc. num.	Length	Score	cTP	CS score	cTP length
I7HUF2	600	0.528	Y	7.795	34
B9VS06	595	0.560	Y	1.817	45
Q9FRX6	562	0.449	−	−0.450	54
P43311	607	0.581	Y	6.236	51
Q9MB14	588	0.564	Y	6.023	50
G0ZAL2	590	0.571	Y	5.525	46
P43309	593	0.575	Y	4.680	47
Q948S3	593	0.577	Y	4.680	47
Q08307	587	0.580	Y	4.524	50
Q08296	585	0.571	Y	5.067	49
Q06355	588	0.574	Y	6.072	47
F1DBB9	584	0.572	Y	6.212	45
O49912	592	0.574	Y	4.751	50
R4JDV1	590	0.576	Y	−1.792	48
C0LU17	603	0.571	Y	5.901	55
Q6UIL3	584	0.545	Y	4.512	32
D1KEC6	556	0.482	−	4.136	51
Q6YHK5	577	0.543	Y	6.853	51
Q06215	606	0.552	Y	3.698	47
P43310	639	0.538	Y	8.027	44

Length is the length of the total amino acid sequence, Score is output score on which the cTP/non-cTP assignment is based (the higher the score, the more certain is that this sequence contains an N-terminal chloroplast transit peptide), cTP is the chloroplast transit peptide (Y means yes, − means no), CS score is the cleavage site score (is defined so that the predicted cleavage site is directly N-terminal of the highest scoring residue within the 40 residues), cTP length gives the predicted length of the chloroplast transit peptide.

2.6. Sequence homologies within PPOs

Phylogenetic analysis was performed using Archaeopteryx 0.9813 A1ST (Han & Zmasek, 2009), and the result is shown in Fig. 4. The amino acid sequences of the organisms involved are listed in Table 3. The phylogenetic tree shows that PPOs cluster in the three groups of plants (blue (black and

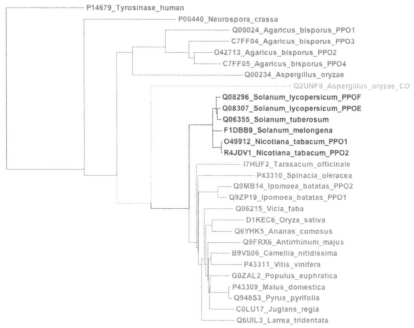

Figure 4 Phylogenetic tree of polyphenol oxidases. The tree was constructed by Archaeopteryx: Visualization, Analysis, and Editing of Phylogenetic Trees (Han & Zmasek, 2009). The length of the sections indicates the relative distances between the sequences. In red (gray in print version), human tyrosinase is shown, in dark green (gray in print version), fungal tyrosinases are shown, in light green (light gray in print version), fungal catechol oxidase is shown, in dark blue (black in print version), plant catechol oxidases are shown, in light blue (light gray in print version), plant catechol oxidases, tyrosinases, and aureusidin synthase are shown (the first six characters in the labels on the figure are the Uniprot database accession numbers, see Table 3).

light gray in print version) branch), fungi (green (gray and light gray in print version) branch), and the human tyrosinase (red (gray in print version) branch). Plant PPOs spread in one group of *Nicotiana* sp. and *Solanum* sp. (dark blue (black in print version) branch, sequence alignment see Fig. 5) and a second big group of all other plant PPOs (light blue (light gray in print version) branch). The dark blue (black in print version) branch includes only CO, but the light blue (light gray in print version) branch includes different PPOs such as plant CO, plant tyrosinase, and aureusidin synthase.

Sequences alignments of the total amino acid sequences (core region, C-terminal domain, and transit peptide) of CO from the dark blue (black in print version) branch (Uniprot numbers Q08296, Q08307, Q06355, F1DBB9, O49912, R4JDV1) show very high sequence identities of

Table 3 Enzymes, their source and Uniprot accession numbers included for the phylogenetic tree

UniProt accession number	Source	Enzyme
I7HUF2	*Taraxacum officinale*	Catechol oxidase
B9VS06	*Camellia nitidissima*	Catechol oxidase
Q9FRX6	*Antirrhinum majus*	Aureusidin synthase
P43311	*Vitis vinifera*	Tyrosinase
Q9MB14	*Ipomoea batatas*	Catechol oxidase
Q9ZP19	*Ipomoea batatas*	Catechol oxidase
G0ZAL2	*Populus euphratica*	Catechol oxidase
P43309	*Malus domestica*	Tyrosinase
Q948S3	*Pyrus pyrifolia*	Catechol oxidase
Q08307	*Solanum lycopersicum*	Catechol oxidase
Q08296	*S. lycopersicum*	Catechol oxidase
Q06355	*Solanum tuberosum*	Catechol oxidase
F1DBB9	*Solanum melongena*	Catechol oxidase
O49912	*Nicotiana tabacum*	Catechol oxidase
R4JDV1	*N. tabacum*	Catechol oxidase
C0LU17	*Juglans regia*	Tyrosinase
Q6UIL3	*Larrea tridentata*	(+)-Larreatricin hydroxylase
D1KEC6	*Oryza sativa*	Catechol oxidase
Q6YHK5	*Ananas comosus*	Catechol oxidase
Q00024	*Agaricus bisporus* PPO1	Tyrosinase
O42713	*A. bisporus* PPO2	Tyrosinase
C7FF05	*A. bisporus* PPO4	Tyrosinase
C7FF04	*A. bisporus* PPO3	Tyrosinase
Q00234	*Aspergillus oryzae*	Tyrosinase
Q2UNF9	*A. oryzae*	Catechol oxidase
P14679	*Tyrosinase human*	Tyrosinase
Q06215	*Vicia faba*	Catechol oxidase
P43310	*Spinacia oleracea*	Catechol oxidase
P00440	*Neurospora crassa*	Tyrosinase

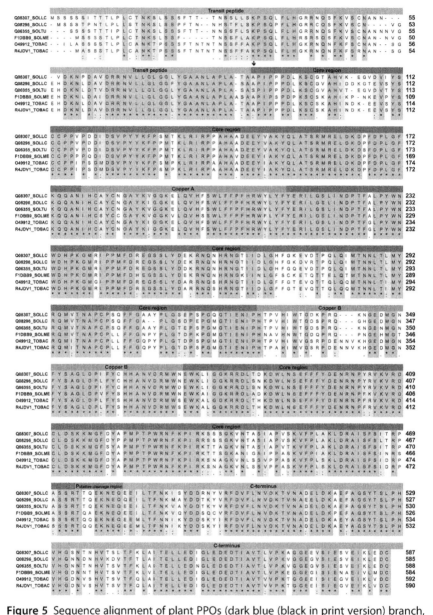

Figure 5 Sequence alignment of plant PPOs (dark blue (black in print version) branch, see Fig. 4).

69.7%. Sequence variation is highest in the transit peptide region of the six CO from *Solanum lycopersicum* (Uniprot numbers Q08296, Q08307), *Solanum tuberosum* (Uniprot number Q06355), *Solanum melongena* (Uniprot number F1DBB9), and *Nicotiana tabacum* (Uniprot number O49912, R4JDV1). Sequence alignments of PPO1 and PPO2 of *N. tabacum* (Uniprot numbers O49912 and R4JDV1) show a sequence identity of 98.8% and the four COs of *Solanum* sp. (Uniprot numbers Q08307, Q08296, Q06355 and F1DBB9) show 75.4% sequence identity.

Fungal tyrosinases (dark green (gray in print version) branch) and the first described fungal AoCO4 (light green (light gray in print version) branch) (Gasparetti et al., 2010; Hakulinen et al., 2013) show very low sequence identities of 2.6% to each other. The four tyrosinase sequences of *A. bisporus* (AbPPO1–4) result in a sequence identity of only 19.0%. In detail, AbPPO1 (Uniprot number Q00024) shows highest sequence identity to AbPPO3 (Uniprot number C7FF04) with 29.2%, and AbPPO2 (Uniprot number O42713) shows highest identity to AbPPO4 (Uniprot number C7FF05) with 57.6%. Tyrosinases from *A. oryzae* (Uniprot number Q00234) and *N. crassa* (Uniprot number P00440) exhibit a sequence identity of 14.8%. The fungal AoCO4 (Uniprot number Q2UNF9) showed sequence identities of 12.5% and 8.5%, respectively, to tyrosinases of *A. oryzae* (Uniprot number Q00234) and *A. bisporus* AbPPO1 (Uniprot number Q00024). Sequence identity of AoCO4 (Uniprot number Q2UNF9) compared with plant PPOs is highest to CO from *Taraxacum officinale* (Uniprot number IHUF2) with 12.8%.

Sequence alignments of the second big group of all other plant PPOs (light blue (light gray in print version) branch), including plant COs, plant tyrosinase, and aureusidin synthase, feature no sequence identity in the transit peptide domain. In the core domain and the C-terminal domain, all previously described conserved motifs of plant PPOs are present. However, total sequence identity over all the sequences of the light blue (light gray in print version) branch is just 13.0%. Aureusidin synthase (Uniprot number Q9FRX6) showed highest sequence identities to CO from *Camellia nitidissima* (Uniprot number B9VS06) with 53.7%, *Popolus euphratica* (Uniprot number G0ZAL2) with 52.1%, *Malus domestica* (Uniprot number P43309) with 50.8%, *Pyrus pyrifolia* (Uniprot number Q948S3) with 49.5%, *V. vinifera* (Uniprot number P43311) with 46.4%, and tyrosinase from *Juglans regia* (Uniprot number COLU17) with 50.8%.

The human tyrosinase (red (gray in print version) branch, Uniprot number P14679) showed highest sequence identity of 10.8% to tyrosinase from *N. crassa* (Uniprot number P00440).

2.7. In vitro activation of PPOs

The latent form of PPOs, consisting of the core domain and the C-terminal domain, can be activated by a variety of treatments using agents as proteases, fatty acids, acid or basic shock, and SDS (Cabanes, Escribano, Gandia-Herrero, Garcia-Carmona, & Jimenez-Atienzar, 2007; Sellés-Marchart, Casado-Vela, & Bru-Martínez, 2007). The use of SDS as an activating agent is particularly interesting, because PPOs are active at high SDS concentrations (1 mM) (Moore & Flurkey, 1990), which would denature or inactivate many other enzymes. It has been suggested that lipids might fulfill the role of SDS and be the physiological counterpart (van Gelder, Flurkey, & Wichers, 1997).

2.8. In vivo activation of PPOs

Activation of tyrosinase from *N. crassa* is processed by a chymotrypsin-like enzyme (Lerch, 1978). Burton, Partis, Wood, and Thurston (1997) showed that a serine protease activates PPOs in *A. bisporus*. In general, it is still not clear when, why, and how the proteolytic cleavage occurs in PPOs and which proteases are responsible for the maturation process.

2.9. Mutants of PPOs

2.9.1 Mutants of plant PPOs

In plant PPOs, first site-directed mutagenesis was performed by Dirks-Hofmeister, Inlow, and Moerschbacher (2012). They mutated Cys197 to serine (C197S) in tetrameric dandelion PPO of *T. officinale* (Uniprot number I7HUF2) and determined its effect on activity and stability. They could prove that C197 forms a disulfide linkage at the surface and stabilizes the tetramer. Kinetic studies showed a decreasing activity of the C197S mutant compared to the wild-type enzyme.

2.9.2 Mutants of fungal PPOs

First site-directed mutagenesis studies on fungal tyrosinase were performed by Nakamura et al. (2000) in *A. oryzae*. They mutated seven histidines, the six copper-coordinating histidines (His63Asn, His84Asn, His93Asn, His290Asn, His294Asn, His332Asn) and the seventh noncopper-coordinating histidine from the $FxxH_{CuB}H$ motif (His333Asn) to asparagine. Furthermore, the cysteine (forming the thioether bridge at the copper A site) was mutated to alanine (Cys82Ala). The authors observed that each mutated amino acid decreased copper binding by approx. 50%, indicating that the mutants contain only approx. 1 g atom of copper/mol of the subunit and concluded that the five mutants, His63Asn, His93Asn, His290Asn, His294Asn, and Cys82Ala, contain only one copper ion, which is fully detectable by EPR. The other three mutants (His84Asn, His332Asn, His333Asn) had no detectable activity, indicating that the mutated residues are essential for activity. In further atomic absorption spectrophotometry experiments, no copper atom content was observed in the seven double mutants His63Asn/His290Asn, His63Asn/His294Asn, His63Asn/His332Asn, His63Asn/His333Asn, Cys82Ala/His290Asn, His84Asn/His333Asn, and His93Asn/His290Asn (Nakamura et al., 2000).

2.9.3 Mutants of bacterial PPOs

Recent research studies on bacterial tyrosinase from *Ralstonia solanacearum* applied random and site-directed mutagenesis (Molloy et al., 2013). By random mutagenesis, two mutants were generated containing four altered amino acids each (RVC10 mutant: T183I/F185Y/N322S/T359M; RV145 mutant: Y119F/V153A/D317Y/L330V), in which the overall catalytic efficiency, K_{cat}/K_m [min^{-1} mM^{-1}], increased 6.7-fold and 16-fold in the two mutants, respectively. For the first time, these mutated amino acids had been identified as amino acids increasing tyrosinase activity. Single mutations of each single amino acid, from random mutagenesis, were performed by site-directed mutagenesis and approved the increase of specific activity. Single mutation variant V153A (from RV145 mutant) exhibited the highest (6.9-fold) improvement in K_{cat} and a 2.4-fold increase in K_m compared to the WT. Two single mutation variants, N322S and T183I (from RVC10 mutant) reduced the K_m up to 2.6-fold for D-tyrosine, but one variant V153A increased the K_m 2.4-fold compared to the wild type.

Kanteev, Goldfeder, Chojnacki, Adir, and Fishman (2013) found by site-directed mutagenesis of tyrosinase from *Bacillus megaterium* (BmTYR) that a major role of the highly conserved Asn205 residue is to stabilize the orientation of the His204 imidazole ring in the binding site, thereby promoting the correct coordination of CuB. Other mutants (as Phe197 to alanine, Met61 and Met184 to leucine) revealed that Phe197, Met61, and Met184, which are located at the entrance to the binding site, are also important for enhancing the diphenolase activity.

3. OXO COMPLEX

First reports on tyrosinase reacting aerobically with molar equivalents of H_2O_2 were published by Jolley, Evans, and Mason (1972). For a review describing the electronic properties of the type-3 copper center, see Solomon et al. (1992). *Met* tyrosinase (the resting form of the enzyme) contains two Cu(II) which are antiferromagnetically coupled. *Met* tyrosinase is reduced by reaction with *o*-diphenols to produce *o*-quinones and the *deoxy* site, which in turn reacts with dioxygen to produce *oxy*tyrosinase, forming a μ-η^2:η^2 bridging mode first described by Kitajima et al. (1989). This can be achieved artificially by the reduction of *met* TYR with hydroxylamine (NH$_2$OH) and the subsequent oxidation with molecular oxygen (O$_2$) (Fujieda, Murata, et al., 2013). *Oxy*tyrosinase hydroxylates monophenols generating *o*-diphenols and the *met* tyrosinase site (Solomon et al., 1992).

3.1. *Oxo* complex investigated by UV/vis absorption spectroscopy

For all type-3 copper centers, the *oxy* form shows a characteristic UV absorption band in the range of 335–350 nm ($\varepsilon = 20{,}000$ M^{-1} cm^{-1}) which is assigned to a peroxo O$_2^{2-}$ $(\pi_\sigma^*) \rightarrow$ Cu(II)$(d_{x^2-y^2})$ CT transition (Eickman et al., 1979; Himmelwright et al., 1980; Solomon et al., 1992, 1994). The second absorption band around 580 nm ($\varepsilon = 1000$ M^{-1} cm^{-1}) is assigned to a peroxo O$_2^{2-}$ $(\pi_v^*) \rightarrow$ Cu(II)$(d_{x^2-y^2})$ CT transition (Eickman et al., 1979; Himmelwright et al., 1980; Solomon et al., 1992, 1994).

3.2. *Oxo* complex investigated by CD spectroscopy

Excitation profiles for the "peroxide" vibration consist of two components and indicate that the absorption and/or CD bands at ~490 and ~570 nm are due to O$_2^{2-}$ $\rightarrow$ Cu(II) CT transitions (Freedman et al., 1976).

3.3. *Oxo* complex investigated by resonance Raman spectroscopy

In resonance Raman spectroscopy, a laser is tuned into an intense absorption band (CT transition), and vibrations associated with the chromophore (i.e., the active site in the protein and the specific ligand–metal bond associated with the CT) become greatly enhanced in Raman scattering intensity (Solomon et al., 2014; Woertink et al., 2009). Raman spectroscopy indicates that the μ-η^2:η^2 side-on peroxo species contains a peroxide moiety with a relatively weak O–O bond (ν_{O-O} ranging between 720 and 760 cm^{-1}) (Baldwin et al., 1992; Solomon et al., 2014).

3.4. *Oxo* complex of catechol oxidase

Investigations on the *oxo* complexes of CO were first described by Eicken et al. (1998), who investigated the two isoforms of sweet potatoes (*I. batatas* Uniprot numbers Q9MB14 and Q9ZP19) in phosphate buffer at pH 6.7 as summarized in Table 4. While for the 39-kDa isoform (Uniprot number Q9ZP19), six equivalents of H_2O_2 are required to reach saturation of the *oxo* complex, only two equivalents are needed for the 40-kDa isoform (Uniprot number Q9MB14). When Gerdemann et al. (2001) applied H_2O_2 as substrate, the 39-kDa isoform showed catalase-like activity, which the 40-kDa isoform did not do. Gerdemann et al. (2001) suggested a mechanism for the catalase activity, where two molecules of hydrogen peroxide bind to the active site. Rompel et al. (1999) studied the saturation of the *oxo* complex of *Lycopus* and *Populus* CO in phosphate buffer at pH 7.0. Saturation was reached with 6 equiv. and 80 equiv. H_2O_2 for *Lycopus* and *Populus*, respectively. A strong resonance Raman peak at 277 cm^{-1}, belonging to a Cu–N (axial His) stretching mode, was described for *oxy* CO from *Populus nigra* (Rompel et al., 1999). CO in lemon balm (*Melissa officinalis*) reaches saturation of the *oxo* complex with 2 equiv. H_2O_2 (Rompel et al., 2012), which is very similar to the 40-kDa CO from sweet potatoes.

Recently, saturation of the *oxo* complex of fungal *A. oryzae* CO was reached with 4–5 equiv. H_2O_2 (Hakulinen et al., 2013). Details about experimental conditions of the *oxo* experiments are summarized in Table 4. CD spectroscopic investigations of CO have not been reported up to now.

3.5. *Oxo* complex of tyrosinase

Resting mushroom tyrosinase had shown bands at 755 and 653 nm in circular dichroism spectra and oxygen-sensitive changes at 350 nm upon treatment with hydroxylamine or hydrogen peroxide (Schoot Uiterkamp, Evans, Jolley, & Mason, 1976).

*Oxy*tyrosinase from *N. crassa* showed an intense Raman peak at 274 cm^{-1}, one defined shoulder at 296 cm^{-1} and a peak at 755 cm^{-1}, indicating that the oxygen is bound as peroxide (Eickman, Solomon, Larrabee, Spiro, & Lerch, 1978).

Studies on the formation of the tyrosinase *oxo* complex were performed on *N. crassa* by Lerch (1976) and Himmelwright et al. (1980) and on *A. bisporus* by Jolley, Evans, Makino & Mason (1974). In *N. crassa* 2 equiv. H_2O_2 were needed to reach saturation, while in *A. bisporus* 3 equiv. were needed. Tyrosinase from walnut leaves (*J. regia*) showed very similar results

Table 4 Studies of forming the oxo complex of PPOs

Source	Enzyme	Size [kDa]	Equivalents H_2O_2	buffer	pH range	$e_{343-345}$ (M^{-1} cm^{-1})	$e_{580-600}$ (M^{-1} cm^{-1})	References
Busycon canaliculatum	Mollusc HC	nd	1	0.2 M potassium phosphate, pH 7	6.0–8.5	nd	nd	Felsenfeld and Printz (1959)
Limulus polyphemus	Arthropod HC	nd	1	0.2 M potassium phosphate, pH 7	6.0–8.5	nd	nd	Felsenfeld and Printz (1959)
Agaricus bisporus	TYR	120	3	26 mM sodium borate, pH 8.6	nd	9000	600	Jolley et al. (1974)
Neurospora crassa	TYR	46	2	10 mM sodium phosphate and 500 mM sodium chloride, pH 6.8	nd	9000	600	Himmelwright et al. (1980) and Lerch, 1976
Ipomoea batatas	CO	39	6	500 mM sodium chloride and 50 mM sodium phosphate, pH 6.7	7.8	6500	450	Eicken et al. (1998)
I. batatas	CO	40	2	500 mM sodium chloride and 50 mM sodium phosphate, pH 6.7	7.8	6500	450	Eicken et al. (1998)
Lycopus europaeus	CO	40	6	50 mM sodium phosphate, pH 7.0	6.5–7.5	3500	1000	Rompel et al. (1999)
Populus nigra	CO	55	80	50 mM sodium phosphate, pH 7.0	6.0–7.0	6000	1000	Rompel et al. (1999)

Melissa officinalis	CO	39	2	50 mM sodium phosphate, pH 7.0	nd	8510	580	Rompel, Büldt-Karentzopoulos, Molitor, and Krebs (2012)
Aspergillus oryzae	CO	44	4–5	10 mM TRIS, pH 7.2	nd	3000	Weak	Hakulinen et al. (2013)
Juglans regia	TYR	39	2	10 mM sodium acetate, pH 5.0	nd	12984	761	Zekiri, Molitor, et al., 2014

nd: not determined.

as CO from *M. officinalis*, the 40-kDa catechol oxidase from *I. batatas* (IbCO), and tyrosinase from *N. crassa* where addition of 2 equiv. H_2O_2 led to saturation (Zekiri, Molitor, et al., 2014).

3.6. *Oxo* complex of aureusidin synthase
Up to date, no investigations have been performed on the *oxo* complex of aureusidin synthase.

3.7. *Oxo* complex of hemocyanin
Resonance Raman investigation of hemocyanin by Freedman et al. (1976) showed a characteristic O–O stretching vibration at 744 cm^{-1} in *Cancer magister* (arthropod) and at 749 cm^{-1} in *Busycon canaliculatum* (mollusc) hemocyanins. *Oxy*hemocyanin shows resonance Raman peaks at 282 cm^{-1} in *C. magister* and 267 cm^{-1} in *B. canaliculatum*. These peaks were not observed in the *deoxy* form and apoprotein and are assigned to Cu–N(imidazole) vibrations of histidine ligands.

The effect of H_2O_2 on arthropod hemocyanin *Limulus polyphemus* and mollusc hemocyanin *B. canaliculatum* was reported by Felsenfeld and Printz (1959). For arthropod *L. polyphemus*, one equivalent of peroxide per mole copper was sufficient to destroy most of the oxygen-carrying capacity of the deoxygenated hemocyanin and it was not possible to regenerate hemocyanin from the attacked material by the use of reducing agents. Mollusc *B. canaliculatum* behaved similarly except that it was possible to regenerate hemocyanin by the use of reducing agents.

4. X-RAY CRYSTALLOGRAPHIC STRUCTURAL DATA OF PPOs AND HEMOCYANINS
4.1. Published structures in the protein data bank
4.1.1 Published structures in the PDB of hemocyanins
The first type-3 copper protein structures that have been published were hemocyanins in the 1990s (among them: Cuff et al. (1998); Magnus, Ton-That, & Carpenter, 1994; Volbeda & Hol, 1989), see Fig. 6A for arthropod hemocyanin of *L. polyphemus* (Magnus et al., 1994) and Fig. 6B for mollusc hemocyanin of *Enteroctopus dofleini* (Cuff et al., 1998).

4.1.2 Published structures in the PDB of catechol oxidases
At the end of the 1990s, the first X-ray structure determination of a type-3 copper enzyme was published (Klabunde et al., 1998). The structure of CO

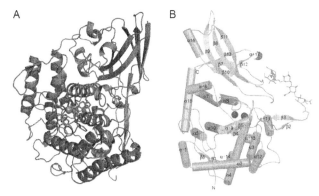

Figure 6 Two exemplary published hemocyanin structures. (A) arthropod hemocyanin of *Limulus polyphemus* (Magnus et al., 1994) (B) mollusc hemocyanin of *Enteroctopus dofleini* (Cuff et al., 1998).

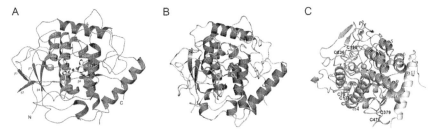

Figure 7 Published catechol oxidase structures. (A) Catechol oxidase of *Ipomoea batatas* (Klabunde et al., 1998), (B) catechol oxidase of *Vitis vinifera* (Virador et al., 2010), and (C) catechol oxidase of *Aspergillus oryzae* (Hakulinen et al., 2013).

is based on the 39-kDa isoform of sweet potatoes (*I. batatas*) (PDB ID: 1BT3, see Fig. 7A). Crystals were obtained of the enzyme in the resting dicupric Cu(II)–Cu(II) state, the reduced dicuprous Cu(I)–Cu(I) form, and in the presence of the inhibitor phenylthiourea (PTU). The catalytic copper center exists in a central four-helix bundle located in a hydrophobic pocket close to the surface. Klabunde et al. (1998) suggested a mechanism for the reaction of IbCO taken the PTU complex into account. The conserved copper binding histidines are placed, with the exception of the second histidine of CuA, on α-helical structure elements. His109 of CuA is covalently linked to Cys92 by a thioether bond. This second histidine is placed on a loop. Also, a bulky phenylalanine residue (in the following called blocker-residue, Table 5) is placed atop the CuA site. The crystal structure of IbCO is depicted in Fig. 7A.

The second CO structure from grapes (*V. vinifera*) has been reported by (Virador et al., 2010) (PDB ID: 2P3X see Fig. 7B). The structures of the

Table 5 Comprehensive structural data list of all published tyrosinases, catechol oxidases, and two exemplary hemocyanins

Type	Organism	PDB ID	Resolution	Quaternary structure	Origin (host)	Full-length or truncated (active/latent)	Size	Active site state	Cu-Cu distance	Crystallization additive or substrate	TE bridge	Placeholder/blocker-residue
TYR	Streptomyces castaneoglobisporus	1WX2	1.8 Å	Dimer of TYR and caddie	rc (Escherichia coli)	fl (A-TYR)	31/13 kDa	Oxy	3.48 Å	H_2O_2 (prepared)	No	Tyr98 (caddie, di-hydroxylated?)/No
TYR	S. castaneoglobisporus	2AHK	1.71 Å	Dimer of TYR and caddie	rc (E. coli)	fl (A-TYR)	31/13 kDa	Met2 ($2 \times H_2O$ bridged)	3.32 Å	$CuSO_4$ (soaked 6 month)	No	Tyr98 (caddie)/No
TYR	S. castaneoglobisporus	2AHL	1.60 Å	Dimer of TYR and caddie	rc (E. coli)	fl (A-TYR)	31/13 kDa	Deoxy ($1 \times H_2O$ bridged)	4.17 Å	NH_2OH (prepared)	No	Tyr98 (caddie)/No
TYR	S. castaneoglobisporus	2ZMX	1.33 Å	Dimer of TYR and caddie	rc (E. coli)	fl (A-TYR)	31/13 kDa	Met ($\sim 2 \times H_2O$ bridged)	3.72 Å	$CuSO_4$ (soaked 36 h)	No	Tyr98 (caddie)/No
TYR	Bacillus megaterium	3NM8	2.00 Å	Homodimer	rc (E. coli)	fl (A-TYR)	34.4 kDa	Met ($1 \times H_2O$ bridged)	3.58 Å	$CuSO_4$ (soaked 36 h)	No	No/Val218
TYR	Aspergillus oryzae	3W6W	1.39 Å	Homodimer	rc (E. coli)	fl (L-TYR)	71.0 kDa	Met2 ($2 \times H_2O$ bridged)	3.58 Å	No	Yes (loop)	Phe513/Val359
TYR	Agaricus bisporus	2Y9W (PPO3)	2.30 Å	Heterotetramer (H_2L_2)	Natural source	C-terminal truncated (A-TYR)	45.3/16.5 kDa	Deoxy ($1 \times H_2O$ bridged)	4.43 Å	$HoCl_3$	Yes (loop)	No/Val283
TYR	A. bisporus	2Y9X (PPO3)	2.78 Å	Heterotetramer (H_2L_2)	Natural source	C-terminal truncated (A-TYR)	45.3/16.5 kDa	Deoxy	4.04-4.36 Å	Tropolone/ $HoCl_3$	Yes (loop)	Tropolone/ Val283

TYR	*A. bisporus*	4OUA (PPO4)	2.76 Å	Heterodimer A-TYR/L-TYR	Natural source	fl and C-terminal truncated (A-TYR/L-TYR)	64.2/43.6 kDa	*Deoxy*	4.4/4.2 Å	TEW	Yes (loop)	Phe454/Ala270
TYR	*Manduca sexta*	3HHS	1.97 Å	Heterodimer (PPO2/PPO1)	Natural source	fl (L-TYR)	80.0/78.9 kDa	*Deoxy* (1 × H_2O bridged in PPO2, zero in PPO1)	4.86/4.53 Å	No	No	Phe88/ Glu395_Phe85/ Ser393
CO	*Ipomoea batatas*	1BT3	2.50 Å	Monomer	Natural source	C-terminal truncated (active)	38.8 kDa	Met (1 × OH bridged)	2.87 Å	No	Yes (helix-loop)	No/Phe261
CO	*I. batatas*	1BT2	2.70 Å	Dimer	Natural source	C-terminal truncated (active)	38.8 kDa	*Deoxy* (1 × OH bridged)	4.37 Å	No	Yes (helix-loop)	No/Phe261
CO	*I. batatas*	1BUG	2.70 Å	Dimer	Natural source	C-terminal truncated (active)	38.8 kDa	*Deoxy*	4.25 Å	Phenylthiourea (PTU)	Yes (helix-loop)	No/Phe261
CO	*Vitis vinifera*	2P3X	2.20 Å	Monomer	Natural source	C-terminal truncated (active)	38.4 kDa	Met (1 × OH bridged)	4.17 Å	No	Yes (helix-loop)	No/Phe259
CO	*Aspergillus oryzea*	4J3P	2.50 Å	Monomer	rc (*Trichoderma reesei*)	fl (active)	42.0 kDa	Oxy (bridged by di-atomic oxygen moiety)	4.23 Å	No	No (helix-loop)	No/Val299

Continued

Table 5 Comprehensive structural data list of all published tyrosinases, catechol oxidases, and two exemplary hemocyanins—cont'd

Type	Organism	PDB ID	Resolution	Quaternary structure	Origin (host)	Full-length or truncated (active/latent)	Size	Active site state	Cu-Cu distance	Crystallization additive or substrate	TE bridge	Placeholder/ blocker-residue
CO	*A. oryzea*	4J3Q	2.90 Å	Homodimer	rc (*T. reesei*)	N-terminal truncated (active)	37.7 kDa	Deoxy ($1 \times H_2O$ bridged)	4.34 Å	No	No (helix–loop)	No/Val299
HC	*Enteroctopus dofleini* (mollusc)	1JS8	2.30 Å	Homodimer	Natural source	fl (latent) part of a gene encoding 2.8 k aa	44.8 kDa	Oxy	3.5 Å	No	Yes (loop)	Leu2830/Leu2689
HC	*Limulus polyphemus* (arthropod)	1JS8	2.40 Å	Monomer	Natural source	fl	72.7 kDa	Oxy	3.6 Å	No	No	Phe49/Thr351

Particularly, structural data that might have an impact on the differences of monophenolase activity and diphenolase activity are listed (e.g., copper distance, active site state, thioether(TE)-bridge presents and location, placeholder (residue that covers CuB and is responsible for latency), blocker (residue that covers CuA and putatively responsible for the lack of diphenolase activity)).

39-kDa sweet potato and CO from grapes are quite similar in overall fold, the location of the helix bundles at the core, and the active site in which three histidines bind each of the two catalytic copper ions, and one of the histidines (His108 from *V. vinifera*, His109 from *I. batatas*) is engaged in a thioether linkage with a cysteine residue (Cys91 from *V. vinifera*, Cys92 from *I. batatas*).

Recently, the first crystal structure of a fungal, extracellular AoCO4 was presented at 2.5 Å resolution (PDB ID: 4J3P and 4J3Q; see Fig. 7C). Hakulinen et al. (2013) described the overall structure as predominantly α-helical. The core of the protein is formed by a four-helix bundle (α-3, α-4, α-8, α-9), and the catalytic copper site is situated within this helical bundle. The geometry of the copper site in the crystal structure of the truncated form of AoCO4 is similar to that of the chemically reduced *deoxy* form of IbCO and the *deoxy* form of *Streptomyces castaneoglobisporus* tyrosinase (ScTYR). The thioether bond found in IbCO and *V. vinifera* is not present in AoCO4. This new class of CO is deficient of the C-terminal active site shielding domain already on the genetic level, however, possesses a proteolytically removable extra α-helix. The enzyme was described to be extracellular located, thus may not need a latent precursor for cell environment protection. Notably, all eukaryotic forms of PPOs except AoCO4 and MsTYR (prophenoloxidase from *Manduca sexta*) possess the thioether bridge at the active site while all prokaryotic ones lack this posttranslational modification (PTM).

4.1.3 Published structures in the PDB of tyrosinases

The first crystal structure of a bacterial tyrosinase from *S. castaneoglobisporus* was solved by Matoba et al. (2006) (PDB ID: 2AHK). They determined the crystal structures of copper-bound and metal-free tyrosinase in a complex with ORF378 designated as a "caddie" protein because it assists with transportation of two Cu(II) ions into the tyrosinase catalytic center. This caddie protein had no similarity with all proteins deposited before into the protein data bank.

With this first tyrosinase structure of a recombinant expressed bacterial tyrosinase, a structural disparity explaining the differing enzymatic active sites was possible. A large vacant space above the active site and a high flexibility of one of the histidine ligands were characterized as the responsible structural features between tyrosinase and CO. This vacant space is occupied by the blocker-residue phenylalanine in CO (IbCO).

The authors also reported that CuA or both copper ions exhibit some flexibility during catalyzes.

The crystal structure of an active, bacterial BmTYR was determined by Sendovski et al. (2011) with a resolution of 2.0–2.3 Å (PDB ID: 3NN8, see Fig. 8A). The structure of a second recombinant expressed bacterial tyrosinase gave new insights in substrate binding to the active site and revealed a certain copper plasticity in the active site. The vacant space in ScTYR is occupied with a valine as a blocker residue in BmTYR, however, still being less bulky than the phenylalanine in IbCO and therefore not hindering the monophenolase activity (see crystal structure of IbCO in Fig. 7A).

Interestingly, one year earlier Li et al. (2009) reported the crystal structure of a heterodimer prophenoloxidase from the moth *Manduca sexta* consisting of two homologous polypeptide chains, PPO1 and PPO2, formed in a back-to-back mode (PDB ID: 3HHS, see Fig. 8B). Both PPOs contain one type-3 copper center, but the acidic residue Glu395 in PPO2 may serve as a general base for deprotonation of monophenolic substrates, which is key to the *ortho*-hydroxylase activity of a phenoloxidase. However, the protein was found to be in its latent state because of a bulky phenylalanine residue, originating from an active site protruding loop, which was placed as a substrate placeholder-molecule atop the CuB site. Furthermore, the authors proposed an activation mechanism mediated by two auxiliary proteins, a PPO activating protease (PAP) and a serine protease homologue (SPH).

Ismaya et al. (2011) achieved to purify isoform PPO3 from the commercially available mushroom tyrosinase extracts. PPO3 of *A. bisporus* was crystallized at a resolution of 2.3 Å (PDB ID: 2Y9W, see Fig. 8C), suggesting that the enzyme is in the *deoxy* state, due to the distance of ∼4.5 Å between the two copper ions. They succeeded in crystallizing the enzyme with the co-crystallization agent $HoCl_3$ in its active form. This year the X-ray crystallographic results of latent PPO4 of *A. bisporus* (AbPPO4) to 2.78 Å resolution in the presence of hexa-tungstotellurate(VI) (POM,

Figure 8 Published tyrosinase structures. (A) Tyrosinase of *Bacillus megaterium* (Sendovski et al., 2011), (B) prophenol oxidase of *Manduca sexta* (Li et al., 2009), and (C) tyrosinase of *Agaricus bisporus* PPO3 (Ismaya et al., 2011).

polyoxometalate) as a co-crystallization agent (Mauracher, Molitor, Al-Oweini, et al., 2014a, 2014b) has been reported (PDB ID: 4OUA). Mauracher, Molitor, Al-Oweini, et al. (2014b) presented the simultaneous presence of both forms, active tyrosinase (A-TYR) and latent tyrosinase (L-TYR), within one single-crystal structure, allowing investigations of the transition between these two forms.

Recently, the crystal structure of a fungal protyrosinase was determined of *A. oryzae* by Fujieda, Yabuta, et al. (2013) at a resolution of 1.39 Å (PDB ID: 3W6W). The enzyme was recombinantly expressed and was crystallized in its holo- and apo-form. In the apo-form, no thioether bridge was found. Therefore, the authors suggested a copper incorporation process, involving the conserved $C(X)_nC$-motif ($n=1-3$), that resulted in the two histidine coordinated copper ions (CuA and CuB) and thereby a triggered formation of the thioether bridge. Furthermore, the structure depicted how the C-terminal domain actually shields the active site with a protruding phenylalanine as "placeholder" for phenolic substrates.

The first crystal structure of a plant originated tyrosinase from *J. regia* was reported by Zekiri, Bijelic, et al. (2014). Tyrosinase was isolated from walnut leaves, purified to homogeneity and crystallized. The crystals diffracted to a resolution of 2.39 Å.

4.2. Structural differences of tyrosinases and catechol oxidases

Tyrosinase is an enzyme capable of *ortho*-hydroxylating tyrosine while CO exclusively oxidizes *ortho*-diphenols. There have been many propositions to actually describe a structural difference between a CO and a tyrosinase ever since the first structural data of both became available (Klabunde et al., 1998; Matoba et al., 2006). In Table 5, parameters are summarized that were used to discuss differences. The initial explanation was the present blocker-residue (Phe) in CO placed above the CuA site. This residue is in tyrosinase either not present or substituted by a less bulky residue (Ala or Val). Another explanation was the presence of the thioether bridge. This PTM was not described for tyrosinases (ScTYR, BmTYR) until the elucidation of AbPPO3 structure (Ismaya et al., 2011). Li et al. (2009) proposed that monophenolase activity is either given if no thioether bridge is present or the bridge is located solely on a loop region and not on secondary structure (α-helix) elements. Therefore, a higher flexibility of the CuA site should be present. Also hemocyanin, that can artificially be functionalized as a tyrosinase (Nillius, Jaenicke, Decker, 2008), has either no thioether bridge (mollusc HC) or it is located solely on a loop (arthropod HC). However, both theories were disproved with the recently published structure of AoCO4 (Hakulinen et al., 2013). This

CO exhibits a small blocker residue (Val) and no thioether bridge, however, still does not possess monophenolase activity for L-tyrosine. Notably, AoCO4 is able to hydroxylate aminophenol and guaiacol.

5. CONCLUSIONS AND OUTLOOK

A significant sequence-specific difference between fungal and plant PPOs exists, due to the nonexistence of a transit peptide in fungi and varying motifs in conserved regions. Based on the sequence information of the core domain, it cannot be predicted if a CO, tyrosinase, or aureusidin synthase is present. Therefore, it is not possible to classify a plant PPO into a CO, tyrosinase, or aureusidin synthase only based on its amino acid sequence.

Despite the X-ray structures of type-3 copper proteins and enzymes that have been published in the last 25 years, still many questions are yet not answered. Although several explanations for a structural disparity causing the difference between tyrosinase and CO have been attempted, as well as for the incorporation of the copper ions to the active site, these issues are still not understood. Also the roles of the cocrystallized subunits (ORF378, ORF239342) especially in AbPPO3 are still not clear. Moreover, there is no comprehensive knowledge how and when the tyrosinase maturation process takes place. Neither responsible proteases for the activation, nor why some tyrosinases are even expressed as active forms, are known. Besides, physiological roles for the many isoforms found especially in *A. bisporus* are not yet conceived. Furthermore, X-ray structure of the chalcone-specific aureusidin synthase would help to understand its special physiological role in pigment formation. There is great hope that with more X-ray structures of type-3 copper proteins and spectroscopic investigations these questions will be addressed.

ACKNOWLEDGMENTS
The research was funded by the Austrian Science Fund (FWF): P25217-N28.

REFERENCES
Altunkaya, A., & Gökmen, V. (2008). Effect of various inhibitors on enzymatic browning, antioxidant activity and total phenol content of fresh lettuce (*Lactuca sativa*). *Food Chemistry*, *107*, 1173–1179.

Baldwin, M. J., Root, D. E., Pate, J. E., Fujisawa, K., Kitajima, N., & Solomon, E. I. (1992). Spectroscopic Studies of Side-On Peroxide-Bridged Binuclear Copper(II) Model Complexes of Relevance to oxyhemocyanin and oxytyrosinase. *Journal of the American Chemical Society*, *114*, 10421–10431.

Burton, K. S., Partis, M. D., Wood, D. A., & Thurston, C. F. (1997). Accumulation of serine proteinase in senescent sporophores of the cultivated mushroom, *Agaricus bisporus*. *Mycological Research, 101*, 146–152.

Cabanes, J., Escribano, J., Gandia-Herrero, F., Garcia-Carmona, F., & Jimenez-Atienzar, M. (2007). Partial Purification of Latent Polyphenol Oxidase from Peach (*Prunus persica* L. Cv. Catherina). Molecular Properties and Kinetic Characterization of Soluble and Membrane-Bound forms. *Journal of Agricultural and Food Chemistry, 55*, 10446–10451.

Coetzer, C., Corsini, D., Love, S., Pavek, J., & Tumer, N. (2001). Control of Enzymatic Browning in Potato (*Solanum tuberosum* L.) by Sense and Antisense RNA from Tomato Polyphenol Oxidase. *Journal of Agricultural and Food Chemistry, 49*, 652–657.

Cuff, M. E., Miller, K. I., van Holde, K. E., & Hendrickson, W. A. (1998). Crystal Structure of a Functional Unit from *Octopus* Hemocyanin. *Journal of Molecular Biology, 278*, 855–870.

Decker, H., & Tuczek, F. (2000). Tyrosinase/catecholoxidase activity of hemocyanins: structural basis and molecular mechanism. *Trends in Biochemical Sciences, 25*, 392–397.

Dirks-Hofmeister, M. E., Inlow, J. K., & Moerschbacher, B. M. (2012). Site-directed mutagenesis of a tetrameric dandelion polyphenol oxidase (PPO-6) reveals the site of subunit interaction. *Plant Molecular Biology, 80*, 203–217.

Douwe de Boer, A., & Weisbeek, P. J. (1991). Chloroplast protein topogenesis: import, sorting and assembly. *Biochimica et Biophysica Acta, 1071*, 221–253.

Eicken, C., Zippel, F., Büldt-Karentzopoulos, K., & Krebs, B. (1998). Biochemical and spectroscopic characterization of catechol oxidase from sweet potatoes (*Ipomoea batatas*) containing a type-3 dicopper center. *FEBS Letters, 436*, 293–299.

Eickman, N. C., Himmelwright, R. S., & Solomon, E. I. (1979). Geometric and electronic structure of oxyhemocyanin: Spectral and chemical correlations on *met* apo, half *met*, *met*, and dimer active sites. *Proceedings of the National Academy of Sciences of the United States of America, 76*, 2094–2098.

Eickman, N. C., Solomon, E. I., Larrabee, J. A., Spiro, T. G., & Lerch, K. (1978). Ultraviolet Resonance Raman Study of Oxytyrosinase. Comparison with Oxyhemocyanins. *Journal of the American Chemical Society, 100*, 6529–6531.

Emanuelsson, O., Nielsen, H., Brunak, S., & von Heijne, G. (2000). Predicting Subcellular Localization of Proteins Based on their N-terminal Amino Acid Sequence. *Journal of Molecular Biology, 300*, 1005–1016.

Emanuelsson, O., Nielsen, H., & von Heijne, G. (1999). ChloroP, a neural network-based method for predicting chloroplast transit peptides and their cleavage sites. *Protein Science, 8*, 978–984.

Endo, T., Kawamura, K., & Nakai, M. (1992). The chloroplast-targeting domain of plastocyanin transit peptide can form a helical structure but does not have a high affinity for lipid bilayers. *European Journal of Biochemistry, 207*, 671–675.

Felsenfeld, G., & Printz, M. P. (1959). Specific Reactions of Hydrogen Peroxide with the Active Site of Hemocyanin. The Formation of "Methemocyanin" *Journal of the American Chemical Society, 81*, 6259–6264.

Flurkey, W. H., & Inlow, J. K. (2008). Proteolytic processing of polyphenol oxidase from plants and fungi. *Journal of Inorganic Biochemistry, 102*, 2160–2170.

Freedman, T. B., Loehr, J. S., & Loehr, T. M. (1976). A Resonance Raman Study of the Copper Protein, Hemocyanin. New Evidence for the Structure of the Oxygen-Binding Site. *Journal of the American Chemical Society, 98*, 2809–2815.

Fujieda, N., Murata, M., Yabuta, S., Ikeda, T., Shimokawa, C., Nakamura, Y., et al. (2013). Activation mechanism of *melB* tyrosinase from *Aspergillus oryzae* by acidic treatment. *Journal of Biological Inorganic Chemistry, 18*, 19–26.

Fujieda, N., Yabuta, S., Ikeda, T., Oyama, T., Muraki, N., Kurisu, G., et al. (2013). Crystal Structures of Copper-depleted and Copper-bound Fungal Pro-tyrosinase: Insights into endogenous cysteine-dependent copper incorporation. *Journal of Biological Chemistry*, *288*, 22128–22140.

Fujita, Y., Uraga, Y., & Ichisima, E. (1995). Molecular cloning and nucleotide sequence of the protyrosinase gene, *melO*, from *Aspergillus oryzae* and expression of the gene in yeast cells. *Biochimica et Biophysica Acta*, *1261*, 151–154.

Gasparetti, C., Faccio, G., Arvas, M., Buchert, J., Saloheimo, M., & Kruus, K. (2010). Discovery of a new tyrosinase-like enzyme family lacking a C-terminally processed domain: production and characterization of an *Aspergillus oryzae* catechol oxidase. *Applied Microbiology and Biotechnology*, *86*, 213–226.

Gaykema, W. P. J., Hol, W. G. J., Vereijken, J. M., Soeter, N. M., Bak, H. J., & Beintema, J. J. (1984). 3.2 Å structure of the copper-containing, oxygen-carrying protein *Panulirus interruptus* haemocyanin. *Nature*, *309*, 23–29.

Gerdemann, C., Eicken, C., Magrini, A., Meyer, H. E., Rompel, A., Spener, F., et al. (2001). Isozymes of *Ipomoea batatas* catechol oxidase differ in catalase-like activity. *Biochimica et Biophysica Acta*, *1548*, 94–105.

Hakulinen, N., Gasparetti, C., Kaljunen, H., Kruus, K., & Rouvinen, J. (2013). The crystal structure of an extracellular catechol oxidase from the ascomycete fungus *Aspergillus oryzae*. *Journal of Biological Inorganic Chemistry*, *18*, 917–929.

Halaouli, S., Asther, Mi., Kruus, K., Guo, L., Hamdi, M., Sigoillot, J.-C., et al. (2005). Characterization of a new tyrosinase from *Pycnoporus* species with high potential for food technological applications. *Journal of Applied Microbiology*, *98*, 332–343.

Halaouli, S., Asther, M., Sigoillot, J.-C., Hamdi, M., & Lomascolo, A. (2006). Fungal tyrosinases: new prospects in molecular characteristics, bioengineering and biotechnological applications. *Journal of Applied Microbiology*, *100*, 219–232.

Han, M. V., & Zmasek, C. M. (2009). phyloXML: XML for evolutionary biology and comparative genomics. *BMC Bioinformatics*, *10*, 356.

Hazes, B., Magnus, K. A., Bonaventura, C., Bonaventura, J., Dauter, Z., Kalk, K. H., et al. (1993). Crystal structure of deoxygenated *Limulus polyphemus* subunit II hemocyanin at 2.18 Å resolution: Clues for a mechanism for allosteric regulation. *Protein Science*, *2*, 597–619.

Himmelwright, R. S., Eickman, N. C., LuBien, C. D., Lerch, K., & Solomon, E. I. (1980). Chemical and Spectroscopic Studies of the Binuclear Copper Active Site of *Neurospora* tyrosinase: Comparison to Hemocyanins. *Journal of the American Chemical Society*, *102*, 7339–7344.

Ismaya, W. T., Rozeboom, H. J., Weijn, A., Mes, J. J., Fusetti, F., Wichers, H. J., et al. (2011). Crystal Structure of *Agaricus bisporus* Mushroom Tyrosinase: Identity of the Tetramer Subunits and Interaction with Tropolone. *Biochemistry*, *50*, 5477–5486.

Jaenicke, E., Büchler, K., Decker, H., Markl, J., & Schröder, G. F. (2011). The Refined Structure of Functional Unit h of Keyhole Limpet Hemocyanin (KLH1-h) Reveals Disulfide Bridges. *IUBMB Life*, *63*, 183–187.

Jolley, R. L., Jr., Evans, L. H., & Mason, H. S. (1972). Reversible oxygenation of tyrosinase. *Biochemical and Biophysical Research Communications*, *46*, 878–884.

Jolley, R. L., Evans, L. H., Makino, N., & Mason, H. S. (1974). Oxytyrosinase. *The Journal of Biological Chemistry*, *249*, 335–345.

Kaintz, C., Molitor, C., Thill, J., Kampatsikas, I., Michael, C., Halbwirth, H., et al. (2014). Cloning functional expression in *E. coli* of a polyphenol oxidase transcript from *Coreopsis grandiflora* involved in aurone formation. *FEBS Letters*, *588*, 3417–3426.

Kanteev, M., Goldfeder, M., Chojnacki, M., Adir, N., & Fishman, A. (2013). The mechanism of copper uptake by tyrosinase from *Bacillus megaterium*. *JBIC, Journal of Biological Inorganic Chemistry*, *18*, 895–903.

Kawamura-Konishi, Y., Tsuji, M., Hatana, S., Asanuma, M., Kakuta, D., Kawano, T., et al. (2007). Purification, Characterization, and Molecular Cloning of Tyrosinase from *Pholiota nameko*. *Bioscience, Biotechnology, and Biochemistry*, *71*, 1752–1760.

Kitajima, N., Fujisawa, K., Moro-oka, Y., & Toriumi, K. (1989). μ-η^2:η^2-Peroxo Binuclear Copper Complex, [Cu(HB(3,5-iPr$_2$pz)$_3$)]$_2$(O$_2$). *Journal of the American Chemical Society, 111*, 8975–8976.

Klabunde, T., Eicken, C., Sacchettini, J. C., & Krebs, B. (1998). Crystal structure of a plant catechol oxidase containing a dicopper center. *Nature Structural Biology, 5*, 1084–1090.

Kupper, U., Niedermann, D. M., Travaglini, G., & Lerch, K. (1989). Isolation and Characterization of the Tyrosinase Gene from *Neurospora crassa*. *The Journal of Biological Chemistry, 264*, 17250–17258.

Lerch, K. (1976). *Neurospora* tyrosinase: Molecular weight, copper content and spectral properties. *FEBS Letters, 69*, 157–160.

Lerch, K. (1978). Amino acid sequence of tyrosinase from *Neurospora crassa*. *Proceedings of the National Academy of Sciences of the United States of America, 75*, 3635–3639.

Li, Y., Wang, Y., Jiang, H., & Deng, J. (2009). Crystal structure of *Manduca sexta* prophenoloxidase provides insights into the mechanism of type 3 copper enzymes. *Proceedings of the National Academy of Sciences of the United States of America, 106*, 17002–17006.

Linzen, B., Soeter, N. M., Riggs, A. F., Schneider, H. J., Schartau, W., Moore, M. D., et al. (1985). The structure of arthropod hemocyanins. *Science, 229*, 519–524.

Magnus, K. A., Ton-That, H., & Carpenter, J. E. (1994). Recent Structural Work on the Oxygen Transport Protein Hemocyanin. *Chemical Reviews, 94*, 727–735.

Martinez, M. V., & Whitaker, J. R. (1995). The biochemistry and control of enzymatic browning. *Trends in Food Science & Technology, 6*, 195–200.

Marusek, C. M., Trobaugh, N. M., Flurkey, W. H., & Inlow, J. K. (2006). Comparative analysis of polyphenol oxidase from plant and fungal species. *Journal of Inorganic Biochemistry, 100*, 108–123.

Matoba, Y., Kumagai, T., Yamamoto, A., Yoshitsu, H., & Sugiyama, M. (2006). Crystallographic Evidence That the Dinuclear Copper Center of Tyrosinase Is Flexible during Catalysis. *Journal of Biological Chemistry, 281*, 8981–8990.

Mauracher, S. G., Molitor, C., Al-Oweini, R., Kortz, U., & Rompel, A. (2014a). Crystallization and preliminary X-ray crystallographic analysis of latent isoform PPO4 mushroom (*Agaricus bisporus*) tyrosinase. *Acta Crystallographica Section F, 70*, 263–266.

Mauracher, S. G., Molitor, C., Al-Oweini, R., Kortz, U., & Rompel, A. (2014b). Latent and active abPPO4 mushroom tyrosinase cocrystallized with hexatungstotellurate(VI) in a single crystal. *Acta crystallographica, D70*, 2301–2315.

Mauracher, S. G., Molitor, C., Michael, C., Kragl, M., Rizzi, A., & Rompel, A. (2014). High level protein-purification allows the unambiguous polypeptide determination of latent isoform PPO4 of mushroom tyrosinase. *Phytochemistry, 99*, 14–25.

Mayer, A. M. (2006). Polyphenol oxidases in plants and fungi: Going places? A review. *Phytochemistry, 67*, 2318–2331.

Mayer, A. M., & Harel, E. (1979). Polyphenol oxidases in plants. *Phytochemistry, 18*, 193–215.

Molloy, S., Nikodinovic-Runic, J., Martin, L. B., Hartmann, H., Solano, F., Decker, H., et al. (2013). Engineering of a Bacterial Tyrosinase for Improved Catalytic Efficiency Towards D-Tyrosine Using Random and Site Directed Mutagenesis Approaches. *Biotechnology and Bioengineering, 110*, 1849–1857.

Moore, B. M., & Flurkey, W. H. (1990). Sodium Dodecyl Sulfate Activation of a Plant Polyphenoloxidase. Effect of sodium dodecyl sulfate on enzymatic and physical characteristics of purified broad bean polyphenoloxidase. *The Journal of Biological Chemistry, 265*, 4982–4988.

Nakamura, M., Nakajima, T., Ohba, Y., Yamauchi, S., Lee, B. R., & Ichishima, E. (2000). Identification of copper ligands in *Aspergillus oryzae* tyrosinase by site-directed mutagenesis. *The Biochemical Journal, 350*, 537–545.

Nakayama, T. (2002). Enzymology of Aurone Biosynthesis. *Journal of Bioscience and Bioengineering, 94*, 487–491.
Nakayama, T., Sato, T., Fukui, Y., Yonekura-Sakakibara, K., Hayashi, H., Tanaka, Y., et al. (2001). Specificity analysis and mechanism of aurone synthesis catalyzed by aureusidin synthase, a polyphenol oxidase homolog responsible for flower coloration. *FEBS Letters, 499*, 107–111.
Nakayama, T., Yonekura-Sakakibara, K., Sato, T., Kikuchi, S., Fukui, Y., Fukuchi-Mizutani, M., et al. (2000). Aureusidin Synthase: A Polyphenol Oxidase Homolog Responsible for Flower Coloration. *Science, 290*, 1163–1166.
Nillius, D., Jaenicke, E., & Decker, H. (2008). Switch between tyrosinase and catecholoxidase activity of scorpion hemocyanin by allosteric effectors. *FEBS Letters, 582*, 749–754.
Queiroz, C., da Silva, A. J. R., Lopes, M. L. M., Fialho, E., & Valente-Mesquita, V. L. (2011). Polyphenol oxidase activity, phenolic acid composition and browning in cashew apple (*Anacardium occidentale*, L.) after processing. *Food Chemistry, 125*, 128–132.
Robinson, N. J., & Winge, D. R. (2010). Copper Metallochaperones. *Annual Review of Biochemistry, 79*, 537–562.
Rompel, A., Büldt-Karentzopoulos, K., Molitor, C., & Krebs, B. (2012). Purification and spectroscopic studies on catechol oxidase from lemon balm (*Melissa officinalis*). *Phytochemistry, 81*, 19–23.
Rompel, A., Fischer, H., Meiwes, D., Büldt-Karentzopoulos, K., Dillinger, R., Tuczek, F., et al. (1999). Purification and spectroscopic studies on catechol oxidases from *Lycopus europaeus* and *Populus nigra*: Evidence for a dinuclear copper center of type 3 and spectroscopic similarities to tyrosinase and hemocyanin. *JBIC, Journal of Biological Inorganic Chemistry, 4*, 56–63.
Schoot Uiterkamp, A. J., Evans, L. H., Jolley, R. L., & Mason, H. S. (1976). Absorption and circular dichroism spectra of different forms of mushroom tyrosinase. *Biochimica et Biophysica Acta, 453*, 200–204.
Sellés-Marchart, S., Casado-Vela, J., & Bru-Martínez, R. (2007). Effect of detergents, trypsin and unsaturated fatty acids on latent loquat fruit polyphenol oxidase: Basis for the enzyme's activity regulation. *Archives of Biochemistry and Biophysics, 464*, 295–305.
Sendovski, M., Kanteev, M., Ben-Yosef, V. S., Adir, N., & Fishman, A. (2011). First Structures of an Active Bacterial Tyrosinase Reveal Copper Plasticity. *Journal of Molecular Biology, 405*, 227–237.
Solomon, E. I., Baldwin, M. J., & Lowery, M. D. (1992). Electronic Structures of Active Sites in Copper Proteins: Contributions to Reactivity. *Chemical Reviews, 92*, 521–542.
Solomon, E. I., Heppner, D. E., Johnston, E. M., Ginsbach, J. W., Cirera, J., Qayyum, M., et al. (2014). Copper Active Sites in Biology. *Chemical Reviews, 114*, 3659–3853.
Solomon, E. I., Tuczek, F., Root, D. E., & Brown, C. A. (1994). Spectroscopy of Binuclear Dioxygen Complexes. *Chemical Reviews, 94*, 827–856.
Spagna, G., Barbagallo, R. N., Chisari, M., & Branca, F. (2005). Characterization of a Tomato Polyphenol Oxidase and Its Role in Browning and Lycopene Content. *Journal of Agricultural and Food Chemistry, 53*, 2032–2038.
Strack, D., & Schliemann, W. (2001). Bifunctional Polyphenol Oxidases: Novel Functions in Plant Pigment Biosynthesis. *Angewandte Chemie International Edition in English, 40*, 3791–3794.
Tessari, I., Bisaglia, M., Valle, F., Samorì, B., Bergantino, E., Mammi, S., et al. (2008). The Reaction of α-Synuclein with Tyrosinase: Possible Implications for Parkinson Disease. *Journal of Biological Chemistry, 283*, 16808–16817.
Tran, L. T., Taylor, J. S., & Constabel, C. P. (2012). The polyphenol oxidase gene family in land plants: Lineage-specific duplication and expansion. *BMC Genomics, 13*, 395.

van Gelder, C. W. G., Flurkey, W. H., & Wichers, H. J. (1997). Sequence and structural features of plant and fungal tyrosinases. *Phytochemistry*, *45*, 1309–1323.

Virador, V. M., Reyes Grajeda, J. P., Blanco-Labra, A., Mendiola-Olaya, E., Smith, G. M., Moreno, A., et al. (2010). Cloning, Sequencing, Purification, and Crystal Structure of Grenache (*Vitis vinifera*) Polyphenol Oxidase. *Journal of Agricultural and Food Chemistry*, *58*, 1189–1201.

Volbeda, A., & Hol, W. G. J. (1989). Crystal structure of hexameric haemocyanin from *Panulirus interruptus* refined at 3.2 Å resolution. *Journal of Molecular Biology*, *209*, 249–279.

Weijn, A., Bastiaan-Net, S., Wichers, H. J., & Mes, J. J. (2013). Melanin biosynthesis pathway in *Agaricus bisporus* mushrooms. *Fungal Genetics and Biology*, *55*, 42–53.

Wichers, H. J., Gerritsen, Y. A. M., & Chapelon, C. G. J. (1996). Tyrosinase isoforms from the fruitbodies of *Agaricus bisporus*. *Phytochemistry*, *43*, 333–337.

Wichers, H. J., Recourt, K., Hendriks, M., Ebbelaar, C. E. M., Biancone, G., Hoeberichts, F. A., et al. (2003). Cloning, expression and characterisation of two tyrosinase cDNAs from *Agaricus bisporus*. *Applied Microbiology and Biotechnology*, *61*, 336–341.

Woertink, J. S., Smeets, P. J., Groothaert, M. H., Vance, M. A., Sels, B. F., Schoonheydt, R. A., et al. (2009). A $[Cu_2O]^{2+}$ core in Cu-ZSM-5, the active site in the oxidation of methane to methanol. *Proceedings of the National Academy of Sciences of the United States of America*, *106*, 18908–18913.

Wu, J., Chen, H., Gao, J., Liu, X., Cheng, W., & Ma, X. (2010). Cloning, characterization and expression of two new polyphenol oxidase cDNAs from *Agaricus bisporus*. *Biotechnology Letters*, *32*, 1439–1447.

Yoruk, R., & Marshall, M. R. (2003). Physicochemical properties and function of plant polyphenol oxidase: A review. *Journal of Food Biochemistry*, *27*, 361–422.

Zekiri, F., Bijelic, A., Molitor, C., & Rompel, A. (2014). Crystallization and preliminary X-ray crystallographic analysis of polyphenol oxidase from *Juglans regia* (*jr*PPO1). *Acta Crystallographica Section F*, *70*, 832–834.

Zekiri, F., Molitor, C., Mauracher, S. G., Michael, C., Mayer, R. L., Gerner, C., et al. (2014). Purification and characterization of tyrosinase from walnut leaves (*Juglans regia*). *Phytochemistry*, *101*, 5–15.

CHAPTER TWO

Biophysical Studies of Matrix Metalloproteinase/Triple-Helix Complexes

Gregg B. Fields[1]

Torrey Pines Institute for Molecular Studies, Port St. Lucie, Florida, USA
[1]Corresponding author: e-mail address: gfields@tpims.org

Contents

1. MMPs and Collagen Hydrolysis   37
2. Structures of Full-Length, Collagenolytic MMPs in Solution and in the Solid State   38
3. Structural Evaluation of MMP Interactions with Collagen   41
4. Mechanism of Collagenolysis   43
5. Heterogeneity in MMP Structures   46
Acknowledgment   47
References   47

Abstract

Several members of the zinc-dependent matrix metalloproteinase (MMP) family catalyze collagen degradation. The structures of MMPs, in solution and solid state and in the presence and absence of triple-helical collagen models, have been assessed by NMR spectroscopy, small-angle X-ray scattering, and X-ray crystallography. Structures observed in solution exhibit flexibility between the MMP catalytic (CAT) and hemopexin-like (HPX) domains, while solid-state structures are relatively compact. Evaluation of the maximum occurrence (MO) of MMP-1 conformations in solution found that, for all the high MO conformations, the CAT and HPX domains are not in tight contact, and the residues of the HPX domain reported to be responsible for the binding to the collagen triple-helix are solvent exposed. A mechanism for collagenolysis has been developed based on analysis of MMP solution structures. Information obtained from solid-state structures has proven valuable for analyzing specific contacts between MMPs and the collagen triple-helix.

1. MMPs AND COLLAGEN HYDROLYSIS

Matrix metalloproteinases (MMPs) are a family of zinc-dependent proteolytic enzymes. Several members of the MMP family are capable of catalyzing the hydrolysis of triple-helical, interstitial (types I–III) collagen

(Fields, 2013). Prior to collagen catabolism, MMPs bind to numerous regions within the collagen triple-helix (Sun, Smith, Hasty, & Yokota, 2000). MMPs then progressively move on collagen fibrils (Saffarian, Collier, Marmer, Elson, & Goldberg, 2004). The mechanism by which MMPs catabolize collagen has been postulated for decades. Recent structural studies of MMPs, in both solution and the solid state, have shed significant light on the enzyme conformations that may participate in collagenolysis. In turn, these structural studies have been utilized to develop models for the stepwise degradation of collagen. Some controversy has arisen due to contradictions between the stepwise models. The contradictions may well be based on the different conditions by which MMP structures were obtained.

2. STRUCTURES OF FULL-LENGTH, COLLAGENOLYTIC MMPs IN SOLUTION AND IN THE SOLID STATE

Full-length MMP-1 structures have been obtained by X-ray crystallographic and NMR spectroscopic analyses. Initially, the X-ray crystallographic structure of full-length, active porcine MMP-1 complexed with N-[3-(N'-hydroxycarboxamido)-2-(2-methylpropyl)-propananoyl]-O-methyl-L-tyrosine-N-methylamide (CIC) was solved (Li et al., 1995). Subsequently, the structure of full-length, activated human MMP-1 Glu219Ala mutant was obtained (pdb: 2CLT) (Iyer, Visse, Nagase, & Acharya, 2006). The activated human MMP-1 structure was highly similar to the structure of porcine full-length MMP-1, with an average root mean square deviation of ∼1.4 Å (Iyer et al., 2006).

The full-length active MMP-1 structure included the N-terminal catalytic (CAT) domain, the linker region, and the C-terminal hemopexin-like (HPX) domain (Fig. 1) (Iyer et al., 2006; Li et al., 1995). The CAT domain possesses five β-strands and three α-helices (Fig. 1). The active site within the CAT domain is too small (∼5 Å) to accommodate the collagen triple-helix (∼15 Å) (Li et al., 1995). The HPX domain is a four-bladed β-propeller, where the blades are notated as bI, bII, bIII, and bIV (Fig. 1).

X-ray crystallographic structures have also been solved for full-length proMMP-1 and MMP-13. Comparison of proMMP-1 and activated MMP-1 indicated that activated human MMP-1 had significant movement in the HPX domain compared with proMMP-1. The magnitude of this movement was exemplified by Phe308, where the displacement was 16 Å (Iyer et al., 2006). The movement of the HPX domain toward the CAT domain widened the cleft between the domains on the active site face of

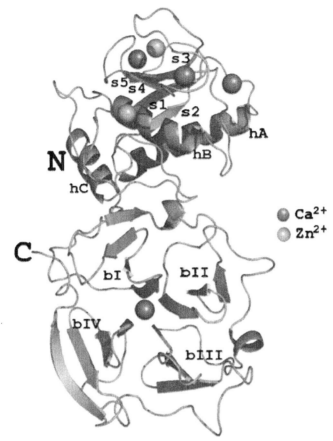

Figure 1 Ribbon representation of the three-dimensional structure of activated, full-length human MMP-1 Glu219Ala mutant. Helices have been colored pink (dark gray in the print version) and the strands shown in green (dark gray in the print version). There are four calcium ions and two zinc ions found in the structure that have been colored gray and orange (light gray in the print version), respectively. The secondary structural elements have been annotated: helices (hA–hC), strands (s1–s5) of the CAT domain, and blades (bI–bIV) of the HPX domain. *Reproduced from Iyer et al. (2006) by the permission of Elsevier.*

the enzyme (Iyer et al., 2006). It was proposed that Arg300 functioned as a pivot for this displacement (Iyer et al., 2006).

The structure of activated full-length human MMP-13 Glu223Ala mutant showed some distinct differences from activated human MMP-1 (Stura, Visse, Cuniasse, Dive, & Nagase, 2013). Most notably, the HPX domain was rotated by ~30° and translated compared with the MMP-1 HPX domain. The displacement was due to changes in the interface between the CAT and HPX domains. The residues Pro236 and Leu322 in MMP-13, replacing Ile and Phe, respectively, in MMP-1, allowed the MMP-13 CAT and HPX domains to move closer together than in MMP-1. Analysis of peptide interactions within the MMP-13 CAT domain implicated the presence of substrate-dependent secondary binding sites (exosites) (Stura et al., 2013).

The radii of gyration (R_g) of the activated human MMP-1 crystallographic structure (2CLT) was 25.7 Å, whereas experimentally determined values from small-angle X-ray scattering (SAXS) data yielded $R_g = 28.5–29.0$ Å (Arnold et al., 2011; Bertini et al., 2009). The X-ray structures were thus more compact than the average solution conformation. In similar fashion, SAXS and single-molecule atomic force microscopy (AFM) analysis of proMMP-9 (Rosenblum et al., 2007) indicated an elongated ellipsoid structure, with $R_g = 50 \pm 2.7$ Å. The domains were separated by $\sim$30 Å.

Full-length active human MMP-1 Glu219Ala mutant was observed by NMR spectroscopy and SAXS to experience a sizable interdomain flexibility and an open-closed equilibrium (Fig. 2) (Arnold et al., 2011; Bertini et al., 2009). Paramagnetic NMR spectroscopic and SAXS data were subsequently used to calculate the maximum occurrence (MO) of conformations of MMP-1 in solution (Cerofolini et al., 2013).

Many of the MMP-1 conformations with the highest MO value were found to have interdomain orientations and positions that could be grouped into a cluster. The structures with the highest MO ($>$35%) had R_g of 29 ± 1.3 Å, further demonstrating that the X-ray structures were more compact than the average solution conformation. Furthermore, the relative orientations of the HPX and CAT domains in the structures with the highest MO were different from those in the X-ray crystallographic structures.

The MO values obtained for the X-ray structures 1SU3 (proMMP-1) and 2CLT (activated MMP-1) were 20% and 19%, respectively. Thus,

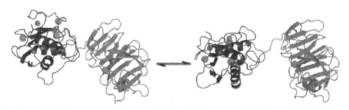

Figure 2 Closed (left) and open/extended (right) forms of full-length MMP-1 in equilibrium. *Reproduced from Bertini et al. (2012) by the permission of the American Chemical Society.*

the compact arrangements of the MMP-1 CAT and HPX domains observed in the X-ray crystallographic structures are not fully representative of the conformations sampled by the protein in solution.

3. STRUCTURAL EVALUATION OF MMP INTERACTIONS WITH COLLAGEN

Unique regulatory sites have been predicted for all members of the MMP family (Udi et al., 2013). A prior study on hydrolysis of collagen by MMP-1, MMP-1 mutants, and general proteases led to the proposal that the collagen triple-helix binds to exosites in the MMP and is then locally unwound/destabilized to allow entry of a single strand into the active site (Chung et al., 2004). Binding sites for the triple-helix within the MMP-1 HPX domain have been identified in solution using hydrogen–deuterium exchange mass spectrometry (HDX-MS) (Lauer-Fields et al., 2009) and NMR spectroscopy (Arnold et al., 2011; Bertini et al., 2012) and in the solid state by X-ray crystallography (Manka et al., 2012). In all cases, triple-helical peptide (THP) models were complexed with MMP-1.

Initially, using HDX-MS and mutational analysis, in MMP-1, Ile290 and Arg291 in the HPX domain A–B loop of blade I were identified as key residues in collagenolysis (Lauer-Fields et al., 2009). Subsequently, NMR spectroscopic studies implicated HPX domain Phe301, Val319, and Asp338 in collagen binding (Fig. 3) (Arnold et al., 2011). Based on the X-ray crystallographic structure of an MMP-1/THP complex (pdb: 4AUO), Phe320 was found to be an important contributor, along with Ile290 and Arg291, to the S_{10}'-binding pocket (Fig. 4) (Manka et al., 2012). The S_{10}'-binding pocket bound the P_{10}' subsite of the THP, which possessed a Leu residue important for the interaction of triple-helices with MMP-1 (Arnold et al., 2011; Manka et al., 2012; Robichaud, Steffensen, & Fields, 2011). In MMP-13, the S_{10}' pocket is shifted, with Arg297 at one edge of the pocket moved 7.3 Å relative to MMP-1 (Stura et al., 2013).

MMP-1 Phe301 was identified as a binding site by NMR spectroscopy (Arnold et al., 2011) but was deemed as buried in the CAT domain/HPX domain interface by X-ray crystallography (Manka et al., 2012). However, in the crystal structure of the MMP-1/THP complex, the THP bound to a closed form of MMP-1 (more closed than 2CLT), and the preferred collagen cleavage site was not correctly positioned for hydrolysis (Manka et al., 2012). This leaves some question as to the physiological relevance of the crystal structure of the MMP-1/THP complex (see later). Phe301 probably

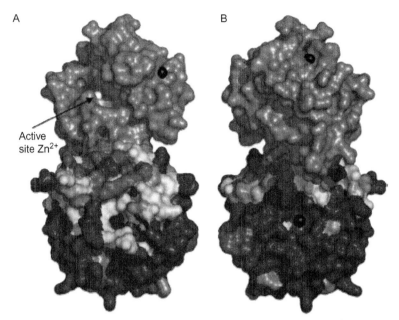

Figure 3 THP recognition surface on the MMP-1 HPX domain as identified by NMR spectroscopy. The HPX domain residues are colored according to the line broadening data from Arnold et al. (2011). The colors range from blue (dark gray in the print version) to cyan (white in the print version), green (light gray in the print version), yellow (white in the print version), and red (light gray in the print version) over the relative intensity range of 40–20%. In panel (B), the HPX domain has been rotated through 180° about the y-axis. *Reproduced from Arnold et al. (2011) by the permission of the American Society for Biochemistry and Molecular Biology.*

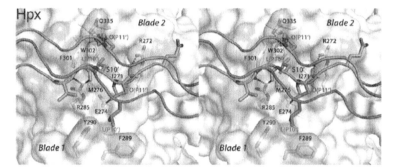

Figure 4 X-ray crystallographic structure of the MMP-1(E219A)/THP complex. Stereo-view of the interactions of the THP chains with the HPX domain. Selected residues making enzyme–substrate contacts are labeled with respective colors and shown in stick representation (N, dark blue, gray in the print version; O, red, dark gray in the print version). The residue numbering does not include the 19-amino acid signal sequence. *Reproduced from Manka et al. (2012) by the permission of the National Academy of Sciences, USA.*

interacts with the triple-helix initially, but then is utilized for domain interaction during collagenolysis (Arnold et al., 2011). Other residues within the HPX domain may also participate in collagen binding (Fig. 4) (Arnold et al., 2011; Lauer-Fields et al., 2009; Manka et al., 2012). In general, the interactions between the triple-helix and the HPX domain are a combination of polar and apolar contacts and are mediated mainly through the 2T (middle) strand of the THP (Manka et al., 2012).

X-ray crystallographic studies of MMP-12 and MMP-8 identified contacts from a single-stranded collagen-like peptide after hydrolysis (Bertini et al., 2006). The peptide fragments Pro-Gln-Gly and Ile-Ala-Gly, representing the cleavage site in the α1(I) collagen chain by collagenolytic MMPs (Fields, 2013), were held in place in the active site (Fig. 5). The N-terminal Pro-Gln-Gly fragment had only a few stabilizing interactions. The carboxylate termini of the Gly residue bound to the active site zinc. A water molecule was also semicoordinated to the zinc and hydrogen bonded to the Glu219 residue that functioned in catalytic activity. As expected, the side chain of the Ile residue from the C-terminal Ile-Ala-Gly fragment entered the S_1' subsite of the enzyme. The peptide was stabilized by four hydrogen bonds with the backbone of the enzyme (carbonyl oxygen atom of Ile and the nitrogen atom of Leu181, nitrogen atom of Ala and the carbonyl oxygen atom of Pro238, carbonyl oxygen atom of Ala and the nitrogen atom of Tyr240, and the nitrogen atom of Gly and the carbonyl oxygen atom of Gly179). The N-terminal nitrogen atom of the Ile-Ala-Gly fragment was hydrogen bonded to the zinc-coordinated water molecule and one oxygen atom of Glu219. The Pro-Gln-Gly fragment left the active site cavity first. The Ile-Ala-Gly fragment underwent a rearrangement in the active site prior to release (Fig. 5).

4. MECHANISM OF COLLAGENOLYSIS

The initial interaction of MMP-1 with collagen is controversial and depends upon which structure is favored by MMP-1 in solution prior to binding the substrate (Bertini et al., 2012; Manka et al., 2012). The highest MO conformations sampled by MMP-1 when free in solution appeared to be much more poised for interaction with collagen than the compact X-ray crystallographic structures (Cerofolini et al., 2013). In the conformations belonging to this cluster (i) the collagen-binding residues of the HPX domain were solvent exposed and (ii) the CAT domain was already correctly positioned for its subsequent interaction with the collagen (Fig. 6).

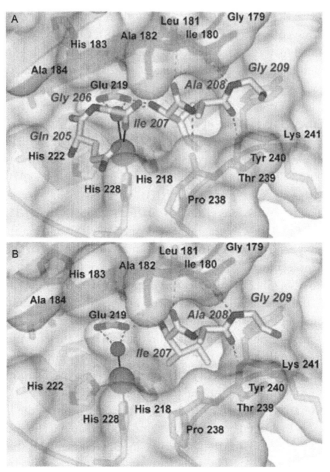

Figure 5 (A) Two-peptide intermediate observed upon soaking active uninhibited MMP-12 crystals with the collagen fragment Pro-Gln-Gly-Ile-Ala-Gly. (B) Ile-Ala-Gly adduct of MMP-12. *Reproduced from Bertini et al. (2006) by the permission of Wiley-VCH Verlag GmbH & Co.*

A structural rearrangement involving a ∼50° rotation around a single axis of the CAT domain with respect to the HPX domain was sufficient to position the CAT domain right in front of the preferred cleavage site in triple-helical collagen. The transition seemed to be feasible at physiological temperature as the difference in free energy between these steps in the pathway was favorable (−0.133 kcal/mol).

The interaction of MMP-1 with a THP has been investigated utilizing NMR spectroscopy, leading to a plausible multistep mechanism for collagenolysis (Bertini et al., 2012). In this mechanism, the initial binding of

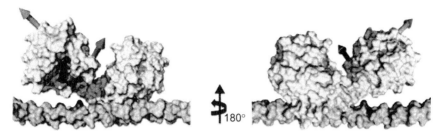

Figure 6 MMP-1 solution structure with the highest MO, where the structure in the right column was rotated 180° about the vertical axis with respect to the left column. Yellow (white in the print version) is the surface representation of MMP-1, blue (black in the print version) is the MMP consensus sequence HEXXHXXGXXH, and orange (light gray in the print version) is the MMP-1 catalytic Zn^{2+}. The blue (dark gray in the print version) and red (gray in the print version) arrows indicate the directions of the helices hA and hC, respectively. *Reproduced from Cerofolini et al. (2013) by the permission of the American Society for Biochemistry and Molecular Biology.*

the HPX domain to the collagen triple-helix through specific residues in blades I and II of the HPX domain is followed by the interaction of the CAT domain with the triple-helix in front of the cleavage site, and by a subsequent back-rotation of the CAT and HPX domains toward the closed conformation observed by X-ray crystallography that drives the unwinding/perturbation of the triple-helix and causes the displacement of one peptide chain into the active site (Fig. 7). The MMP-1/THP complex in which one strand of the triple-helix is displaced from the other two (Fig. 7B) was strongly supported by changes in intensity of the interhelical NOEs upon THP binding (Bertini et al., 2012). Hydrolysis of the first strand is presumably followed by rapid hydrolysis of the other two strands.

The binding of the MMP-1 CAT domain to the THP in front of the Gly-Ile sequence of chain 1T to be cleaved (Fig. 7A) is enthalpically favored and entropically disfavored. The isolated CAT domain has negligible affinity for the THP (Bertini et al., 2012), gaining the ability to bind once held in place by the HPX domain in MMP-1. The proposed back-rotation of MMP-1 toward the X-ray closed structure and the associated liberation of chain 1T from the compact THP structure (Fig. 7B) imply a balance of enthalpic gains (the closing of MMP-1 and the establishment of productive contacts between chain 1T and the active site of MMP-1) and losses (the breaking of the H bonds between chain 1T and the other two THP chains). The cleavage step (Fig. 7C) is energetically favorable (Welgus, Jeffrey, & Eisen, 1981).

The action of MMP-9 on triple-helical collagen fragments, as observed by AFM (Rosenblum et al., 2010), reflects similar behavior to that proposed

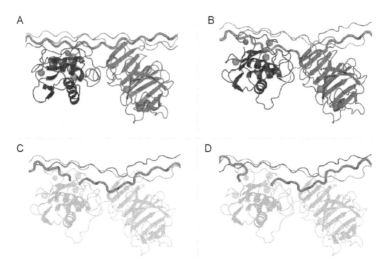

Figure 7 The initial steps of collagenolysis. (A) The extended full-length MMP-1 binds THP chains 1T and 2T (the leading and middle strands, respectively) at Val23-Leu26 with the HPX domain and the residues around the cleavage site with the CAT domain. The THP is still in a compact conformation. (B) Closed full-length MMP-1 interacting with the released 1T chain (in magenta, dark gray in the print version). (C) After hydrolysis, both peptide fragments (C- and N-terminal) are initially bound to the active site. (D) The C-terminal region of the N-terminal peptide fragment is released. *Reproduced from Bertini et al. (2012) by the permission of the American Chemical Society.*

for MMP-1 hydrolysis of collagen. Analysis of MMP-9 lobe-to-lobe distance showed that upon binding of collagen the enzyme goes from an elongated conformation in solution (lobe-to-lobe distance of 74 ± 10 Å) to a more globular conformation upon binding (lobe-to-lobe distance of 60 ± 20 Å). The conformational change was associated with denaturation of the triple-helix.

Models for collagenolysis have also been proposed based on X-ray crystallographic structures (Manka et al., 2012; Stura et al., 2013). In these models, significant movement of the triple-helix (3–6 Å), rotation/sliding of the triple-helix, and/or significant movement of a single strand from the triple-helix (~4 Å) is needed to accommodate a chain in the active site. It is not clear if these models are energetically favorable.

5. HETEROGENEITY IN MMP STRUCTURES

X-ray crystallographic and NMR spectroscopic structures of MMP-12 CAT bound to several inhibitors have been compared (Bertini et al.,

2005). The structures were nonidentical, with many loops having dynamic behavior on a variety of timescales. Even with the same inhibitor bound to the S_1' subsite, the 245–248 region showed conformational heterogeneity. In general, different structures in the solid state (X-ray crystal) corresponded to mobility in solution (NMR). Analysis of other MMPs showed similar disorder in loop regions. The authors ultimately stated, "Flexibility/conformational heterogeneity in crucial parts of the CAT domain is a rule rather than an exception in MMPs, and its extent may be underestimated by inspection of one x-ray structure" (Bertini et al., 2005).

While X-ray crystallographic analysis of an MMP-1/THP complex has revealed binding of the THP to a closed form of MMP-1, it has been noted that the preferred collagen cleavage site was not correctly positioned for hydrolysis (Manka et al., 2012). The flexibility of MMP-1 domains, and particularly the highly favored extended conformation, has a critical role in enzyme function. The closed structures observed by X-ray crystallography appear to be most relevant for collagenolysis once the collagen has bound and the triple-helix is destabilized.

ACKNOWLEDGMENT

MMP studies in the Fields Laboratory have been supported by the National Institutes of Health (CA98799, AR063795, and NHLBI contract 268201000036C-0-0-1).

REFERENCES

Arnold, L. H., Butt, L., Prior, S. H., Read, C., Fields, G. B., & Pickford, A. R. (2011). The interface between catalytic and hemopexin domains in matrix metalloproteinase 1 conceals a collagen binding exosite. *Journal of Biological Chemistry, 286*, 45073–45082.

Bertini, I., Calderone, V., Cosenza, M., Fragai, M., Lee, Y.-M., Luchinat, C., et al. (2005). Conformational variability of matrix metalloproteinases: Beyond a single 3D structure. *Proceedings of the National Academy of Sciences of the United States of America, 102*, 5334–5339.

Bertini, I., Calderone, V., Fragai, M., Luchinat, C., Maletta, M., & Yeo, K. J. (2006). Snapshots of the reaction mechanism of matrix metalloproteinases. *Angewandte Chemie International Edition in English, 45*, 7952–7955.

Bertini, I., Fragai, M., Luchinat, C., Melikian, M., Mylonas, E., Sarti, N., et al. (2009). Interdomain flexibility in full-length matrix metalloproteinase-1 (MMP-1). *Journal of Biological Chemistry, 284*, 12821–12828.

Bertini, I., Fragai, F., Luchinat, C., Melikian, M., Toccafondi, M., Lauer, J. L., et al. (2012). Structural basis for matrix metalloproteinase 1 catalyzed collagenolysis. *Journal of the American Chemical Society, 134*, 2100–2110.

Cerofolini, L., Fields, G. B., Fragai, M., Geraldes, C. F. G. C., Luchinat, C., Parigi, G., et al. (2013). Examination of matrix metalloproteinase-1 (MMP-1) in solution: A preference for the pre-collagenolysis state. *Journal of Biological Chemistry, 288*, 30659–30671.

Chung, L., Dinakarpandian, D., Yoshida, N., Lauer-Fields, J. L., Fields, G. B., Visse, R., et al. (2004). Collagenase unwinds triple helical collagen prior to peptide bond hydrolysis. *EMBO Journal, 23*, 3020–3030.
Fields, G. B. (2013). Interstitial collagen catabolism. *Journal of Biological Chemistry, 288*, 8785–8793.
Iyer, S., Visse, R., Nagase, H., & Acharya, K. R. (2006). Crystal structure of an active form of human MMP-1. *Journal of Molecular Biology, 362*, 78–88.
Lauer-Fields, J. L., Chalmers, M. J., Busby, S. A., Minond, D., Griffin, P. R., & Fields, G. B. (2009). Identification of specific hemopexin-like domain residues that facilitate matrix metalloproteinase collagenolytic activity. *Journal of Biological Chemistry, 284*, 24017–24024.
Li, J., Brick, P., O'Hare, M. C., Skarzynski, T., Lloyd, L. F., Curry, V. A., et al. (1995). Structure of full-length porcine synovial collagenase reveals a C-terminal domain containing a calcium-linked, four bladed b-propeller. *Structure, 15*, 541–549.
Manka, S. W., Carafoli, F., Visse, R., Bihan, D., Raynal, N., Farndale, R. W., et al. (2012). Structural insights into triple-helical collagen cleavage by matrix metalloproteinase 1. *Proceedings of the National Academy of Sciences of the United States of America, 109*, 12461–12466.
Robichaud, T. K., Steffensen, B., & Fields, G. B. (2011). Exosite interactions impact matrix metalloproteinase collagen specificities. *Journal of Biological Chemistry, 286*, 37535–37542.
Rosenblum, G., Van den Steen, P. E., Cohen, S. R., Bitler, A., Brand, D. D., Opdenakker, G., et al. (2010). Direct visualization of protease action on collagen triple helical structure. *PLoS One, 5*, e11043.
Rosenblum, G., Van den Steen, P. E., Cohen, S. R., Grossmann, J. G., Frenkel, J., Sertchook, R., et al. (2007). Insights into the structure and domain flexibility of full-length pro-matrix metalloproteinase-9/gelatinase B. *Structure, 15*, 1227–1236.
Saffarian, S., Collier, I. E., Marmer, B. L., Elson, E. L., & Goldberg, G. (2004). Interstitial collagenase is a Brownian rachet driven by proteolysis of collagen. *Science, 306*, 108–111.
Stura, E. A., Visse, R., Cuniasse, P., Dive, V., & Nagase, H. (2013). Crystal structure of full-length human collagenase 3 (MMP-13) with peptides in the active site defines exosites in the catalytic domain. *FASEB Journal, 27*, 4395–4405.
Sun, H. B., Smith, G. N., Jr., Hasty, K. A., & Yokota, H. (2000). Atomic force microscopy-based detection of binding and cleavage site of matrix metalloproteinase on individual type II collagen helices. *Analytical Biochemistry, 283*, 153–158.
Udi, Y., Fragai, M., Grossman, M., Mitternacht, S., Arad-Yellin, R., Calderone, V., et al. (2013). Unraveling hidden regulatory sites in structurally homologous metalloproteases. *Journal of Molecular Biology, 425*, 2330–2346.
Welgus, H. G., Jeffrey, J. J., & Eisen, A. Z. (1981). Human skin fibroblast collagenase: Assessment of activation energy and deuterium isotope effect with collagenous substrates. *Journal of Biological Chemistry, 256*, 9516–9521.

CHAPTER THREE

Catalytic Mechanisms of Metallohydrolases Containing Two Metal Ions

Nataša Mitić[*,1], Manfredi Miraula[*,†], Christopher Selleck[†], Kieran S. Hadler[†], Elena Uribe[‡], Marcelo M. Pedroso[†], Gerhard Schenk[†,1]

[*]Department of Chemistry, National University of Ireland, Maynooth, Maynooth, Co. Kildare, Ireland
[†]School of Chemistry and Molecular Biosciences, The University of Queensland, Brisbane, Queensland, Australia
[‡]Department of Biochemistry and Molecular Biology, University of Concepción, Concepción, Chile
[1]Corresponding authors: e-mail address: natasa.mitic@nuim.ie; schenk@uq.edu.au

Contents

1. Introduction 50
2. Metallo-β-Lactamases, Major Culprits in the Emergence of Antibiotic Resistance 50
3. Methionine Aminopeptidase, a Target for Novel Anticancer Drugs 57
4. Glycerophosphodiesterase, a Very Promiscuous Potential Bioremediator 60
5. PAP, an Alternative Target to Treat Osteoporosis 65
6. Conclusions and Outlook: Agmatinase, an Emerging Target for Biotechnological Applications 69
Acknowledgments 71
References 72

Abstract

At least one-third of enzymes contain metal ions as cofactors necessary for a diverse range of catalytic activities. In the case of polymetallic enzymes (i.e., two or more metal ions involved in catalysis), the presence of two (or more) closely spaced metal ions gives an additional advantage in terms of (i) charge delocalisation, (ii) smaller activation barriers, (iii) the ability to bind larger substrates, (iv) enhanced electrostatic activation of substrates, and (v) decreased transition-state energies. Among this group of proteins, enzymes that catalyze the hydrolysis of ester and amide bonds form a very prominent family, the metallohydrolases. These enzymes are involved in a multitude of biological functions, and an increasing number of them gain attention for translational research in medicine and biotechnology. Their functional versatility and catalytic proficiency are largely due to the presence of metal ions in their active sites. In this chapter, we thus discuss and compare the reaction mechanisms of several closely related enzymes with a view to highlighting the functional diversity bestowed upon them by their metal ion cofactors.

1. INTRODUCTION

At least one-third of all known enzymes contain metal ion cofactors as an essential requirement for catalytic activity. The roles of the metal ion(s) in the active site are diverse and include electron and oxygen transfer, electrophilic catalysis, substrate binding, and the activation of metal ion-bound nucleophiles (Meyers, 1996). In the case of polynuclear metalloenzymes, the presence of two (or more) closely spaced metal ions gives an additional advantage in terms of (i) charge delocalization, (ii) smaller activation barriers (i.e., decreased transition-state energies), (iii) the ability to bind larger substrates, (iv) enhanced electrostatic activation of substrates, and (v) more facile and effective activation of catalytically relevant nucleophiles (Dismukes, 1996). While the majority of metalloenzymes can easily be categorized as being either mono-, bi-, or even polynuclear, a number of enzymes exist where this classification is less straightforward. In these cases, the number of metal ions employed in the catalytic mechanism may vary in different phases of catalysis or may be dependent on the substrate and/or metal ion involved. A glycerophosphate diesterase from *Enterobacter aerogenes* (GpdQ) is one example of an enzyme that acquires its second metal ion only in the presence of substrate molecules (Schenk et al., 2012). In order to develop a comprehensive understanding of the mechanisms of binuclear enzymes, the following questions need to be addressed: (i) how many metal ion binding sites are present, (ii) how many metal ion binding sites are occupied under normal (or physiological) conditions, and (iii) how many metal ions are required to achieve maximal activity and can this change depending on the substrate? Presented here is an examination of a selection of enzymes for which answers to some of these questions have been discussed. In our treatise, for illustrative purposes, we have concentrated largely (but not exclusively) on systems our groups have been investigating. We thus apologize to all the authors whose contributions are not covered here.

2. METALLO-β-LACTAMASES, MAJOR CULPRITS IN THE EMERGENCE OF ANTIBIOTIC RESISTANCE

β-Lactam antibiotics are the most widely used class of drugs for the treatment of bacterial infections and have been prescribed for over 70 years (Crowder, Spencer, & Vila, 2006; Daumann, Schenk, & Gahan, 2014; Page & Badarau, 2008; Phelan et al., 2014). These antibiotics are divided into four main groups, penicillins, cephalosporins, monobactams, and carbapenems. The characteristic feature of these antibiotics is the presence of a four-membered β-lactam ring. The bactericidal properties of β-lactam antibiotics are afforded by their ability to disrupt the structural integrity of the cell membrane of these pathogens. This occurs by preventing the synthesis of the peptidoglycan layer of cell walls (Crowder et al., 2006). However, bacteria have evolved to resist the action of β-lactam antibiotics, and one of the most efficient mechanisms through which this occurs is by the production of β-lactamases, a family of enzymes that hydrolyze the cyclic amide bond of the antibiotic, rendering the drug ineffective (Crowder et al., 2006). Of the four classes of β-lactamases, three classes (A, C, and D) rely on an active site serine residue for hydrolysis, while class B β-lactamases—or metallo-β-lactamases (MβLs)—employ zinc(II) as cofactor. Our focus here is solely on the latter class, which are further divided into three classes, B1, B2, and B3, depending on their amino acid sequence, substrate specificity, and metal ion requirement (Bebrone, 2007; Garau et al., 2004; Heinz & Adolph, 2004; Page & Badarau, 2008). Active site structures for members of each of the three classes are shown in Fig. 1. More recently, a fourth class has been identified (B4), but no structural information is yet available (Hou et al., 2014; Vella et al., 2013).

The metal ion composition of MβLs and the role(s) of the metal ions in catalysis has been the subject of extensive enquiry and debate, as summarized in some recent reviews (Daumann, Schenk, et al., 2014; Phelan et al., 2014). MβLs possess two Zn(II) binding sites, referred to as the "histidine" or Zn1 site, and "cysteine" or Zn2 site in B1 and B3 MβLs. Members of the B2 subclass only require one Zn(II) ion for catalysis, bound to the "serine"

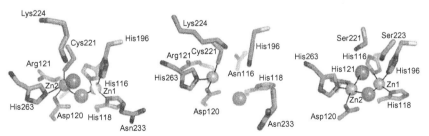

Figure 1 Metallo-β-lactamase active sites. *B. cereus* BcII (B1; left), *A. hydrophila* CphA (B2; center) and *S. maltophilia* L1 (B3; right). Zinc ions are shown as blue spheres and water molecules are shown as red spheres. *The figure was copied with permission from Phelan et al. (2014).*

or Zn2 site (Fig. 1). Binding of a metal ion in the Zn1 site of these MβLs leads to their inactivation. The first available crystal structure of a B1-type MβL was that from the *Bacillus cereus* (i.e., BcII), which revealed the presence of only one zinc ion bound in the Zn1 site (Carfi et al., 1995). This observation indicated that the two metal ion binding sites may have different affinities. It is somewhat surprising that available data for metal ion binding affinities are still scarce, with reported values varying by orders of magnitude depending on methods applied (Phelan et al., 2014). Using fluorescence spectroscopy, binding constants of 0.62 nM for Zn1 and 1.5 μM for Zn2 have been reported (de Seny et al., 2001). Using equilibrium dialysis, the corresponding values obtained were 0.3 and 3 μM, respectively (Paul-Soto et al., 1999). Despite the significant discrepancy, the overall trend supports differential binding affinities of the two metal centers, in agreement with the crystallographic data mentioned above. In the presence of substrate, the binding affinity of the Zn1 site may be enhanced toward the picomolar range, while the affinity of the more weakly bound Zn2 ion is only modestly affected (Wommer et al., 2002). This result has led to the suggestion that BcII, and possibly all B1-type MβLs, operate as mononuclear enzymes *in vivo* where the free zinc concentration is estimated to be in the pico- or femtomolar range (Outten & O'Halloran, 2001; Wommer et al., 2002). Indeed, BcII has been shown to be catalytically active in the mononuclear form. However, the characterization of this mononuclear enzyme has been hindered by the observation that the addition of a single equivalent of Zn(II) to apo-BcII results in a mixture of two monometallic species (de Seny et al., 2001; Hemmingsen et al., 2001; Llarrull, Tioni, Kowalski, Bennett, & Vila, 2007; Paul-Soto et al., 1999). Saturation of BcII with excess metal ions results in a binuclear enzyme with a catalytic activity approximately twice that of the mononuclear species (depending on the substrate being hydrolyzed). In contrast to the BcII enzyme, CcrA from *Bacteroides fragilis*, another B1-type MβL, binds both metal ions very tightly (Crowder, Wang, Franklin, Zovinka, & Benkovic, 1996). Initial kinetic studies reported activity in both the mono- and binuclear forms of CcrA (Paul-Soto et al., 1998). Subsequent investigations determined that only the binuclear enzyme is active and the previously reported activity of a mononuclear form arose in fact from a mixture of apoenzyme and binuclear enzyme (Fast, Wang, & Benkovic, 2001). Thus, while it appears that some MβLs from the B1 subclass may be catalytically active in mononuclear form, the catalytically optimal composition requires a binuclear metal ion center.

Subclass B3 MβLs are functionally similar to their B1 counterparts (despite considerable variation in overall sequence homology) in that they generally exhibit maximal activity in the binuclear form and display a broad substrate specificity (Frere, Galleni, Bush, & Dideberg, 2005). Examples of this subclass include L1 from *Stenotrophomonas maltophilia*, FEZ-1 from *Legionella gormanii*, and BJP-1 from *Bradyrizobium japonicum* (Crowder, Walsh, Banovic, Pettit, & Spencer, 1998; Mercuri et al., 2001; Spencer, Clarke, & Walsh, 2001; Stoczko, Frère, Rossolini, & Docquier, 2006). Reported metal ion binding affinities for these enzymes are in the nM range for both metal ions, consistent with the proposal that these MβLs operate in binuclear form *in vivo* (Crowder et al., 1998; Mercuri et al., 2001; Stoczko et al., 2006). An exception may be the GOB-18 enzyme from *Elizabethkingia meningoseptica*, which appears to be most active in mononuclear form, with Zn(II) bound to the Zn2 site (Morán-Barrio et al., 2007). The discovery of GOB-18 has challenged the notion that all broad spectrum MβLs (from the B1 and B3 subclasses) are most active in their binuclear forms.

MβLs from subclass B2, in contrast, are fully active only when a single metal ion is coordinated to the Zn2 site (Bebrone et al., 2008). Population of the second (Zn1) site leads to noncompetitive inhibition (Hernandez-Valladares et al., 1997). Furthermore, B2-type MβLs are drastically limited in terms of the range of antibiotics they can hydrolyze, with only carbapenems being suitable substrates (Hernandez-Valladares et al., 1997). ImiS from *Aeromonas sobria*, Sfh-1 from *Serratia fonticola*, and CphA from *Aeromonas hydrophila* are examples from the B2 class (Garau et al., 2005; Saavedra et al., 2003; Sharma et al., 2006). A crystal structure of CphA in complex with the substrate biapenem illustrates the coordination of a single metal ion at the Zn2 site (Garau et al., 2005). Subsequent studies have confirmed that this is the only metal ion essential for catalysis. The affinities of Zn(II) for the catalytically relevant Zn2 site and the inhibitory Zn1 site have been estimated to $K_d < 10$ pM and 46 μM, respectively (Hernandez-Valladares et al., 1997; Wommer et al., 2002). Inhibition by binding of Zn(II) to the Zn1 site is believed to be due to the immobilization of the catalytically important His118 and His196 residues, as well as folding of the Gly232-Asn233 loop, leading to a hindered active site (Bebrone, 2007; Bebrone et al., 2009).

Members from the recently identified B4 subclass of MβLs have not yet been studied in great detail. However, a combination of kinetic and spectroscopic measurements indicated that the enzyme SPR-1 from *S. proteamaculans* is mononuclear in its resting state. Addition of a substrate, however, promotes the formation of a catalytically relevant binuclear metal center

(Vella et al., 2013). Insofar, SPR-1 resembles the organophosphate-degrading enzyme GpdQ from *E. aerogenes* (discussed below) (Schenk et al., 2012).

The above discussion illustrated the diversity of various MβLs with respect to their interactions with their metal ion cofactors. Not surprisingly, the proposed reaction mechanisms employed by MβLs are equally flexible and have been shown to be influenced by a variety of factors including the source of the enzyme, the identity of the metal ion(s) bound to the active site and the substrate being turned over (Badarau & Page, 2006; Garau et al., 2005; Hawk et al., 2009; Laraki et al., 1999; Page & Badarau, 2008; Spencer et al., 2001). However, the proposed models can be generally categorized as mono- and binuclear models and their overall features are briefly outlined in the following paragraphs.

A model for the mononuclear reaction mechanism is based on the reaction of BcII from *B. cereus* with the substrate cefotaxim (Carfi et al., 1995; Fabiane et al., 1998), which is proposed to bind to the active site through a series of hydrogen bonds that incorporate the second coordinate sphere residues Asn233 and Lys 224, as well as a water-mediated salt bridge involving the Zn(II)-bound hydroxide ligand (Fig. 2). This Zn(II)-bound hydroxide is oriented in an ideal position for a nucleophilic attack, an orientation that is stabilized by interactions with Asp120, Arg121, Cys221, and His263. Except for Arg121, these amino acids are ligands of Zn2 in B1- and B3-type MβLs (Abriata, Gonzalez, & Llarrull, 2008; de Seny et al., 2001). Electrostatic interactions with these four amino acids, coupled with the Lewis activation imparted by the metal ion, lead to enhancement of the nucleophilicity of the hydroxide ion. Kinetic studies of the pH dependence of the reaction suggest that these factors reduce the pK_a of the hydroxide from 15.7 to 5.6 (Bounaga, Laws, Galleni, & Page, 1998). Substrate hydrolysis is initiated by the attack of this hydroxide on the carbon atom of the cyclic amide carbonyl moiety (Fig. 2). Subsequent rearrangements lead to coordination of a water molecule to Zn(II) to form a penta-coordinate, metastable transition state (Dal Peraro, Vila, Carloni, & Klein, 2007). Due to the spectroscopically silent nature of Zn(II), this transition state has not been characterized experimentally, but is supported by computations (Dal Peraro, Llarrull, & Rothlisberger, 2004). It is proposed that the Zn(II)-bound water molecule acts as a proton shuttle to donate a proton to the nitrogen atom of the cyclic amide ring, thereby inducing cleavage of the C–N bond; this step is believed to be rate-limiting (Ullah et al., 1998).

A model for the hydrolysis of cefotaxime by a binuclear MβL has been proposed largely based on studies with CcrA from *B. fragilis*. Comparison of

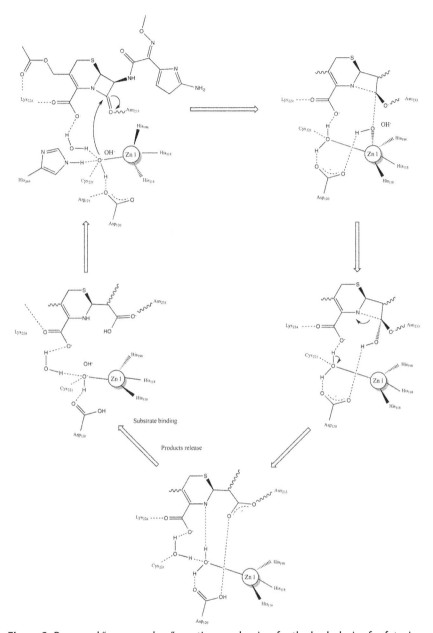

Figure 2 Proposed "mononuclear" reaction mechanism for the hydrolysis of cefotaxime by the *B. cereus* MβL BcII. *Adapted and modified from Dal Peraro, Llarrull, et al. (2004).*

the one- and two-metal ion mechanisms indicate that the incorporation of an additional Zn(II) ion allows hydrolysis to proceed via a more efficient, highly concerted single-step enzymatic reaction, with a lower activation free energy (Dal Peraro, Llarrull, et al., 2004; Dal Peraro, Vila, & Carloni, 2004; Dal Peraro et al., 2007; Estiu, Suarez, & Merz, 2006). In the "binuclear" mechanism, the presence of a second metal ion aids in the orientation of the nucleophilic hydroxide species, which is now bridging the two metal ions. Although this nucleophile experiences fewer electrostatic interactions than in the mononuclear center, the double-Lewis activation more than compensates for this loss, evidenced by the increased reactivity of the binuclear enzyme (Fig. 3). In contrast to the mononuclear mechanism, where the metal ion in the Zn2 site generates both the nucleophile and a water molecule for proton donation in a multistep reaction, the binuclear center of CcrA provides the nucleophile and proton donor concomitantly. Thus, the nucleophilic attack on the C atom, the cleavage of the C–N bond of the cyclic amide ring and the protonation of the nitrogen atom of the product proceed via a single transition state (Fig. 3), which may be the reason for the higher catalytic efficiency observed for binuclear MβLs (Dal Peraro et al., 2007).

MβLs are a family of enzymes for which evolution may be accelerated by development and application of new antibiotics. This has led to the discovery of many varieties of MβLs, displaying a broad diversity in structure, metal ion and substrate binding, and reaction mechanism (Badarau & Page, 2006; Cricco, Orellano, Rasia, Ceccarelli, & Vila, 1999; Dal Peraro et al., 2007; Heinz & Adolph, 2004; Morán-Barrio et al., 2007; Paul-Soto et al., 1999; Sharma et al., 2006). In most cases, the binuclear enzymes appear to have the advantage of a higher catalytic efficiency and lower activation barriers. However, it is still not understood why some B1-type MβLs may operate

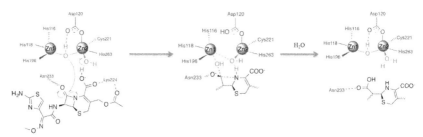

Figure 3 Proposed "binuclear" reaction mechanism for the hydrolysis of cefotaxime by the *B. fragilis* MβL CcrA. *Adapted and modified from Dal Peraro, Llarrull, et al. (2004).*

both in mono- and binuclear form, while B2-type MβLs are inhibited by binding of a second metal ion (in the Zn1 site) (Phelan et al., 2014). Further mechanistic studies are thus essential, especially those relevant for substrate turnover under physiological conditions, to direct the rational design of potent inhibitors as promising drug leads to combat the further spread of antibiotic resistance.

3. METHIONINE AMINOPEPTIDASE, A TARGET FOR NOVEL ANTICANCER DRUGS

Methionine aminopeptidase (MetAP) catalyzes the removal of N-terminal methionine residues from newly synthesized polypeptide chains (Lowther & Matthews, 2002). It is known to play a critical role in the growth of new blood vessels, tumors, and metastasis, and thus is a target for the development of new anticancer agents (Bradshaw & Yi, 2002; Folkman, 1995). However, the design of inhibitors as potential drugs has been hindered by the uncertainty surrounding the number of metal ions that MetAP employs *in vivo*, as well as the identity of these metal ions. Several research groups have sought to clarify these issues, although no general agreement has yet been reached. For yeast MetAP, the high activity of the Zn(II)-derivative in comparison to the Co(II) one has led to the proposal that only Zn(II) may be physiologically important (Walker & Bradshaw, 1998). Based on experiments performed anaerobically, Fe(II) has been suggested as the physiological cofactor for *E. coli* MetAP (D'Souza & Holz, 1999). While a similar conclusion has been reached for MetAP from *Pyrococcus furiosus* the homologous human enzyme is believed to operate using Mn(II) (Ghosh, Grunden, Dunn, Weiss, & Adams, 1998; Meng et al., 2002; Wang et al., 2003). Computational studies have suggested that the lowest activation barrier for catalysis is achieved in the case of di-Zn(II) MetAP (binuclear Co(II), Mn(II), and Fe(II) models display very high energetic barriers to hydrolysis), implicating Zn(II) as the physiologically ideal cofactor for all MetAPs (Leopoldini, Russo, & Toscano, 2007). Despite this ongoing debate regarding the physiological metal ion, MetAPs can be activated *in vitro* using most divalent metal ions, including Co(II), Zn(II), Fe(II), Mn(II), Ni(II), Ca(II), Mg(II), and Cu(II). A number of spectroscopic studies have focussed on the dicobalt derivative, since Co(II) is an excellent probe for the study of active site geometry and chemistry (Ben-Bassat et al., 1987; Chang, Teichert, & Smith, 1992; Evdokimov et al., 2007; Larrabee et al., 1997; Leopoldini et al., 2007; Lowther, Orville, et al., 1999;

Lowther, Zhang, Sampson, Honek, & Matthews, 1999; Walker & Bradshaw, 1998). Many effective inhibitors have been reported for MetAP; however, the potency of these compounds appears to depend heavily on the number and type of metal ion(s) present in the active site (Huang et al., 2007; Li et al., 2003; Ye et al., 2004), highlighting the need for an in-depth understanding of the physiologically relevant reaction mechanism of MetAP.

Crystal structures of MetAP reveal two possible binding sites for metal ions at the active site. A schematic representation of the binuclear Co(II) center in *E. coli* MetAP is shown in Fig. 4. The two Co(II) ions are bridged by a glutamate and aspartate group, both in μ-1,3 fashion, as well as a water (or hydroxide) ligand. The five-coordinate Co1 is coordinated by an additional glutamate and histidine ligand. Metal ion binding affinities for Co1 have been estimated to 0.05–6 μM (D'Souza, Bennett, Copik, & Holz, 2000; D'Souza et al., 2002; Meng et al., 2002). The second metal ion, Co2, is more loosely coordinated by a bidentate aspartate residue and a terminal aqua/hydroxide molecule; its binding affinity is estimated to be ~2.5 mM (D'Souza et al., 2000, 2002; Meng et al., 2002). As a result of this poor binding affinity of the Co2 site, it has been proposed that MetAP may operate as a mononuclear enzyme, a model that is further supported by the finding that apo-MetAP can be nearly fully activated by the addition of a single equivalent of Co(II) or Mn(II) (D'Souza et al., 2000; Ye, Xie, Ma, Huang, & Hanzlik, 2006).

Despite the possibility that the mononuclear form of MetAP may be the biologically relevant one, the majority of functional studies have focused on the binuclear form. Several crystal structures of MetAP, mostly with inhibitors bound to the binuclear metal center, have been reported to date (Douangamath et al., 2004; Hu, Addlagatta, Matthews, & Liu, 2006; Oefner et al., 2003; Schiffmann, Heine, Klebe, & Klein, 2005; Xie et al., 2006). In combination with kinetic and spectroscopic investigations, a model for a binuclear catalytic mechanism for *E. coli* MetAP has been proposed (Fig. 5; Lowther & Matthews, 2002; Lowther, Orville, et al., 1999;

Figure 4 Schematic representation of active site structure of Co-MetAP.

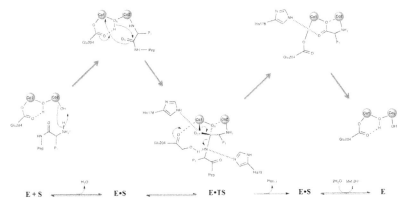

Figure 5 Proposed "binuclear" reaction mechanism for MetAP. *Adapted and modified from Lowther and Matthews (2002).*

Figure 6 Proposed "mononuclear" reaction mechanism for MetAP. *Adapted and modified from Lowther, Orville, et al. (1999).*

Lowther, Zhang, et al., 1999). The initial substrate binding to Co2 leads to a displacement of a water/hydroxide (E + S). Structural rearrangements lead to the activation of the bridging hydroxide nucleophile, which attacks the C atom of the substrate's carbonyl moiety (E•S). The transfer of the proton from the attacking nucleophile to the Co1 ligand Glu204 triggers the dissociation of this amino acid from the metal ion, leading to a transition state where the substrate forms three coordination bonds with the binuclear metal center (E•TS). The enzyme–substrate adduct is stabilized by interactions with His79 and His178. Dissociation of the cleaved peptide results in the formation of L-methionine, which is bound to the metal ions via a μ-1,1-carboxylate bridge; in addition, the amine group of the product binds to Co2 (E•P). Release of the product from the active site enables coordination of solvent molecule to Co1 and Co2 to regenerate the resting state of the enzyme (E).

A crystal structure of a mononuclear Mn(II)-containing MetAP has also been reported (Ye et al., 2006). The metal ion is bound to the site corresponding to the Co1 site above (Fig. 6). The peptide substrate approaches the active site with the carbonyl of the scissile peptide bond and binds to the metal ion. The negatively charged Asp97 and Asp108 residues move closer to the positively charged amino group of the peptide, thus providing charge stabilisation and orienting the substrate in a position optimal for hydrolysis (Chiu, Lee, Lin, Tam, & Lin, 1999; Klinkenberg, Ling, & Chang, 1997). Hydrogen bonding occurs between His79 and the scissile amide nitrogen atom, as well as between His178 and the scissile carbonyl oxygen atom. The metal ion Glu204 acts as a general base to deprotonate the metal ion-bound water to hydroxide, which then acts as the nucleophile in an attack on the carbonyl group of the peptide. This hypothesis is supported by EPR experiments which indicate that both the substrate and nucleophile are bound to the metal ion in the initial stages of catalysis (Copik, Waterson, Swierczek, Bennett, & Holz, 2005). The nucleophilic attack generates a tetrahedral intermediate. The subsequent cleavage of the peptide leaving group is accompanied by the transfer of protons to both Glu204 and His178. In the last step of the catalytic cycle, L-methionine is released and the active site regenerated. Throughout the proposed catalytic cycle the metal ion changes from penta- to hexacoordinate, in agreement with observed structures of MetAP bound to inhibitors (Lowther & Matthews, 2002; Lowther, Zhang, et al., 1999; Ye et al., 2004, 2006). The mononuclear mechanism differs from the binuclear one in that Asp97 and Asp108 replace the second metal ion as "anchor" for the substrate at the catalytic center. Also, in the binuclear form the nucleophile is doubly Lewis activated by both metal ions, whereas in mononuclear MetAP, the nucleophilicity of the terminal hydroxide molecule is enhanced by interactions with Glu204.

A third mechanism has been suggested in which the substrate enhances the affinity of the second metal binding site, thus promoting the formation of a binuclear metal center. This model is supported by NMR experiments (Evdokimov et al., 2007) and has been observed in other binuclear metallohydrolases, in particular the organophosphate-degrading GpdQ from E. aerogenes. Details will be discussed in the following section.

4. GLYCEROPHOSPHODIESTERASE, A VERY PROMISCUOUS POTENTIAL BIOREMEDIATOR

GpdQ from E. aerogenes is a highly promiscuous binuclear metallohydrolase which catalyzes the hydrolysis of all three classes of phosphate ester substrates, i.e., mono-, di-, and triesters (Schenk et al., 2012).

Physiologically, the role of GpdQ is to cleave the 3′-5′ phosphodiester bond of glycerophosphodiesters such as glycerol-3-phosphoethanolamine (Larson, Ehrmann, & Boos, 1983). This process is essential for the regulation of phospholipid remodeling and synthesis (Raetz, 1986). However, beyond its physiological role, GpdQ has been shown to have great potential as an enzymatic bioremediator (Ely et al., 2007). It has been shown to be active toward a range of organophosphate pesticides (triesters) and is capable of degrading EA2192, the highly toxic product of the hydrolysis of VX (one of the most potent nerve agents known) (Ghanem, Li, Xu, & Raushel, 2007). Therefore, GpdQ has promising applications in the detoxification of pesticide-contaminated water and soil, and also in the degradation of nerve agent stockpiles, as required by the Chemical Weapons Convention. A major advancement toward using GpdQ for a bevy of applications is the recent development of a system where the enzyme was attached to the surface of magnetic nanoparticles; activity in this immobilized state was maintained in the excess of 4 months (Daumann, Schenk, & Gahan, 2014).

While the exact metal ion composition of GpdQ is unknown, it is likely to coordinate at least one Fe(II) ion, based on metal ion analysis of the recombinantly expressed enzyme (Jackson et al., 2008). Identification of the second metal ion employed by GpdQ *in vivo* is hindered by the low metal binding affinity at the second site (β site; Fig. 7; Hadler et al., 2008). It is, however, proposed that the Fe(II)–Zn(II) combination may be the most prevalent, based on natural abundance rather than catalytic optimization (Daumann, McCarthy, et al., 2013; Jackson et al., 2008). A study with the heterobinuclear metal ion combinations Fe(II)–Zn(II), Fe(II)–Cd(II), Fe(II)–Mn(II), and Fe(II)–Co(II) *in vitro* showed that the Fe(II) ion has a lower binding affinity to the α site than the other metal ions. Moreover, catalytic efficiencies of these heterobinuclear derivatives were lower than those of their homobinuclear counterparts (Daumann, McCarthy, et al., 2013). Catalytic activity is highest for the binuclear Cd(II) and Mn(II) derivatives; however, activity can be reconstituted with the apoenzyme using virtually any divalent transition metal ion, highlighting the promiscuous nature of the enzyme (Daumann, McCartyh, et al., 2013). Activity, however, cannot be restored using alkaline earth metal ions such as Ca(II) or Mg(II) (Hadler, Mitic, et al., 2010; Pedroso et al., 2014). A number of crystal structures of GpdQ in both the resting state, and with molecules bound to the active site, have been reported (Hadler et al., 2008; Jackson et al., 2007, 2008). The binuclear center is composed of a tightly bound, six-coordinate

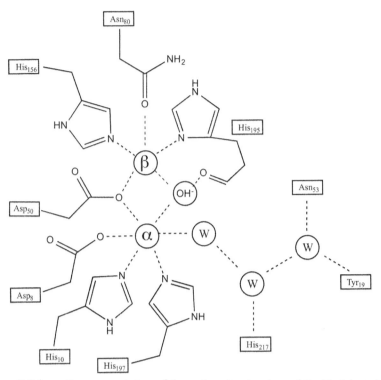

Figure 7 Schematic representation of the active site structure of GpdQ. *Adapted and modified from Hadler et al. (2008).*

α metal ion and a more loosely bound, five-coordinate β metal ion (Fig. 7), bridged by a hydroxide molecule and Asp50 in μ-1,1 mode (Hadler et al., 2008). An additional six amino acid residues complete the primary coordination spheres of the two metal ions. The active site structure of GpdQ is very similar to that of other binuclear phosphoesterases, in particular that of purple acid phosphatases (PAPs) (Guddat et al., 1999; Lindqvist, Johansson, Kaija, Vihko, & Schneider, 1999; Schenk et al., 2008, 2005, 2008; Sträter, Lipscomb, Klabunde, & Krebs, 1996), 5′-nucleotidase (Knöfel & Sträter, 1999), and Mre11 nuclease (Höpfner et al., 2001). PAPs will be discussed in more detail in the next section.

The affinity of Mn(II) for the two binding sites has been measured using EPR spectroscopy, and quantitated to $K_d = 29$ μM (α site) and 344 μM (β site) (Hadler et al., 2009). However, in the presence of the reaction product/substrate analogue phosphate, the affinity for the β metal ion is significantly enhanced (56 μM), while that of the α site remains largely unchanged (Hadler et al., 2009). Thus, it was proposed that GpdQ is predominantly an inactive mononuclear state if no substrate is present, but upon the addition of a substrate the enzyme is activated by the formation of a binuclear metal center in the active site ("substrate-promoted" activation) (Hadler et al., 2008). This unusual active site assembly as well as the catalytic cycle of GpdQ have been studied using site-directed mutagenesis, variable temperature, variable field magnetic circular dichroism (MCD) spectroscopy, as well as steady-state kinetics and stopped-flow fluorescence measurements (Hadler, Mitic, et al., 2010; Hadler et al., 2008).

The following mechanism has been proposed (Hadler, Gahan, et al., 2010; Hadler et al., 2009; Hadler, Mitic, et al., 2010; Hadler et al., 2008). In the resting state, only the α metal ion is bound to GpdQ. MCD experiments indicate the metal ion is six-coordinate, harboring two water ligands (Fig. 8A). Rapid binding of the substrate to the active site, most likely via second coordination sphere interactions, leads to an inactive mononuclear enzyme–substrate complex where the binuclear center is primed for incorporation of the second metal ion (Fig. 8B). Binding of the metal ion in the β site and its subsequent coordination to the substrate (Fig. 8C) leads to an active conformation. Stopped-flow fluorescence experiments indicate that slow structural rearrangements involving the dissociation of the bond between the β metal ion and Asn80 leads to the optimally active form of the enzyme. Thus, GpdQ employs a regulatory reaction mechanism in which reactivity is enhanced as a result of the coordination flexibility of Asn80. Measurements of the pH dependence of the catalytic parameters identify the terminal hydroxide bound to the α metal ion as hydrolysis-initiating nucleophile; its attack on the phosphorus atom of the substrate leads to the release of the alcohol leaving group (Fig. 8D). The phosphate moiety of the substrate remains initially bound to both metal ions in μ-1,3 mode (Fig. 8E), but its gradual dissociation from the active site provides the enzyme with two options. If further substrate molecules are available, the binuclear center remains intact (with Asn80 still uncoordinated) and the addition of a water molecule from the solvent to the α site and substrate coordination to the metal ion in the β site facilitates the next cycle of catalysis (Fig. 8F). However, once the substrate is depleted the metal ion in the β site dissociates from the active site and the enzyme returns to its mononuclear, catalytically inactive resting state (Fig. 8E).

As mentioned earlier, GpdQ displays considerable promise for applications in bioremediation due to its ability to degrade a broad range of phosphate esters. However, the intricate regulatory mechanism outlined in the

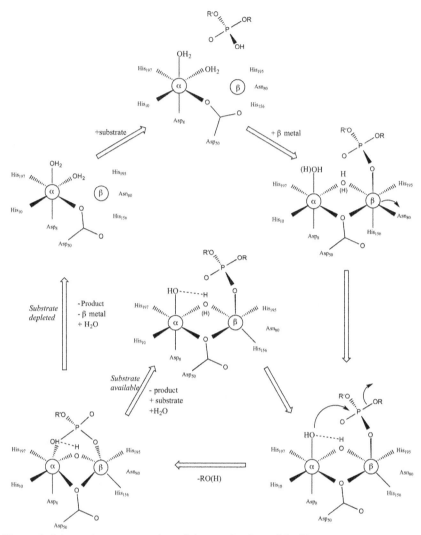

Figure 8 Schematic representation of the mechanism of GpdQ.

preceding paragraph complicates the enzyme's use for such applications, for which a continuously active catalyst is of benefit. In order to achieve this, mutations in the first and second coordination spheres were introduced to probe their effect on the affinity of the metal ions. While the replacement of the flexible ligand Asn80 by an aspartate abolished the coordination flexibility in the β site, this mutation also greatly reduces the catalytic efficiency of the enzyme (Daumann, Comba, et al., 2013; Hadler et al., 2009; Hadler,

Mitic, et al., 2010). More promising are, however, mutations in the second coordination sphere, where the replacement of His81, Ser127, and His217 led to considerable improvements in metal ion binding affinities without affecting the catalytic performance of the enzyme (Daumann, Comba, et al., 2013; Hadler et al., 2009). One of these mutants was successfully attached to magnetic nanoparticles (Daumann, Larrabee, et al., 2014), providing the basis for GpdQ as a useful agent in bioremediation.

In summary, GpdQ employs an intricate, tightly regulated reaction mechanism in which substrate-promoted activation and coordination flexibility play a major role. GpdQ represents an intermediate in the continuum from purely mononuclear to purely binuclear enzymes, in that the reaction cycle employs both the mono- and binuclear forms of the enzyme. The resting enzyme is mononuclear and substrate induces formation of the active binuclear enzyme. It may be that this active site assembly mechanism may play an important physiological regulatory role.

5. PAP, AN ALTERNATIVE TARGET TO TREAT OSTEOPOROSIS

PAPs have been identified in and extracted from various plant, animal, and fungal sources and are likely to occur in only a limited number of microorganisms (Flanagan, Cassady, Schenk, Guddat, & Hume, 2006; Schenk, Guddat, et al., 2000; Schenk, Korsinczky, Hume, Hamilton, & de Jersey, 2000; Schenk et al., 2012; Schenk, Mitic, Hanson, & Comba, 2013). In mammals, PAPs from pig, cow, human, mouse, and rat have been studied in detail (Allen, Nuttleman, Ketcham, & Roberts, 1989; Campbell et al., 1978; Davis, Lin, & Averill, 1981; Ek-Rylander, Bill, Norgard, Nilsson, & Andersson, 1991; Funhoff, Klaassen, Samyn, Van Beeumen, & Averill, 2001; Hayman & Cox, 1994; Hayman, Warburton, Pringle, Coles, & Chambers, 1989; Ketcham, Baumbach, Bazer, & Roberts, 1985; Ketcham, Roberts, Simmen, & Nick, 1989; Ljusberg, Ek-Rylander, & Andersson, 1999; Marshall et al., 1997; Merkx, Pinkse, & Averill, 1999; Mitić et al., 2005). Plant PAPs have been studied from red kidney bean (*Phaseolus vulgaris*) (Beck et al., 1986; Klabunde, Sträter, Fröhlich, Witzel, & Krebs, 1996; Schenk, Elliott, et al., 2008; Schenk, Peralta, et al., 2008), sweet potato (*Ipomoea batatas*) (Durmus, Eicken, Spener, & Krebs, 1999; Schenk et al., 2001; Schenk, Carrington, Hamilton, de Jersey, & Guddat, 1999; Schenk et al., 2005; Schenk, Ge, et al., 1999), soybean (*Glycine max*) (Schenk, Ge, et al., 1999), duckweed (*Spirodela oligorrhiza*) (Nakazato et al., 1998; Nishikoori, Washio, Hase, Morita, & Okuyama, 2001), tomato (*Lycopersicon esculentum*) (Bozzo, Dunn, & Plaxton, 2006; Bozzo, Raghothama, & Plaxton, 2002, 2004), potato (*Solanum tuberosum*) (Zimmermann et al., 2004), yellow lupin (Antonyuk, Olczak, Olczak, Ciuraszkiewicz, & Strange, 2014), and thale cress (*Arabidopsis thaliana*) (del Pozo et al., 1999; Veljanovski, Vanderbeld, Knowles, Snedden, & Plaxton, 2006).

Several crystal structures of PAPs have been described. Despite a low degree of overall amino acid, sequence homology between enzymes from different kingdoms their catalytic sites display a remarkable similarity (Mitić et al., 2006; Schenk et al., 2012, 2013). The two metal centers are coordinated by seven invariant ligands (one aspartate, one tyrosine, and one histidine for Fe^{3+} and two histidines and one asparagine for M^{2+}), with an aspartate residue bridging the two ions (Fig. 9). However, the number of solvent-derived ligands (water molecules) is still subject to debate. In the only known structure of a PAP in the resting state (from red kidney bean), the resolution at 2.9 Å is too low to allow the identification of water ligands (Sträter et al., 1995). The presence of a bridging water ligand has since been observed in several PAP crystal structures (Guddat et al., 1999; Klabunde et al., 1996; Lindqvist et al., 1999; Sträter et al., 2005) and has been

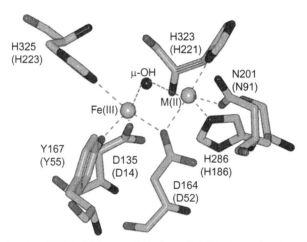

Figure 9 Active site of PAPs. All known PAPs have Fe(III) in the active site which forms a charge-transfer complex with an invariant tyrosine ligand (Y167/55), resulting in the characteristic purple (dark gray in print) color of the enzyme. Residue labels refer to the sequences of red kidney bean PAP and pig PAP (in brackets).

confirmed spectroscopically and magnetochemically (Day et al., 1988; Schenk et al., 2001; Smoukov et al., 2002; Wang & Que, 1998; Wang, Randall, True, & Que, 1996; Yang, McCormick, & Solomon, 1997). An interesting variation is observed in the active site of sweet potato PAP (Schenk et al., 2005), the only known enzyme, to date, with a confirmed Fe(III)–Mn(II) center (Schenk et al., 2001). The crystal structure, in the presence of phosphate, shows that the anion coordinates the two metal ions in an unusual tripodal mode whereby one of the oxygen atoms of the phosphate group bridges the two metal ions. The location of this oxygen atom is equivalent to that of the bridging water ligand in other PAPs. This observation has been interpreted in terms of the bridging oxygen acting as the reaction-initiating nucleophile (Schenk et al., 2005).

A comprehensive model for the reaction mechanism employed by PAP has been proposed, incorporating all the available structural, catalytic, and spectroscopic data (Fig. 10; Schenk, Elliott, et al., 2008; Schenk, Peralta, et al., 2008). In the initial phase of the reaction, a precatalytic complex is formed, where the phosphate group of the substrate does not directly

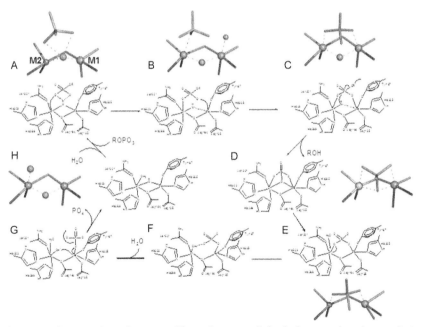

Figure 10 Proposed mechanism of binuclear metallohydrolase-catalyzed esterolysis. *Reproduced with permission from Schenk, Elliott, et al. (2008).*

coordinate to the metal ions (Twitchett et al., 2002). The bridging μ-hydroxide appears to play an essential role in the initial binding and orientation of the substrate (Fig. 10A). In this precatalytic state, both metal ion sites are five-coordinate with distorted trigonal-bipyramidal geometry. The only solvent molecule in the active site is the μ-hydroxide. The formation of the precatalytic complex is followed by a substrate rearrangement that involves firstly a monodentate coordination of an oxygen atom of the phosphate group to the divalent metal ion (Fig. 10B), and secondly the formation of a μ-1,3 substrate complex (Fig. 10C). The resulting μ-1,3 phosphate complex, visualized in the structure of oxidized (inactive) pig PAP (Guddat et al., 1999), places the phosphorus atom in an ideal position for a nucleophilic attack by the μ-hydroxide moiety. Nucleophilic attack by the μ-hydroxide and esterolysis of the substrate (depending on the basicity of the leaving group, these steps may occur in a concerted or sequential manner (Schenk et al., 2005)) leaves the phosphate bound to the active site in a tripodal geometry (Fig. 10D).

It is important to point out that the chemical step may be affected by both the substrates used in the reaction and the metal ion composition of the active site (Cox, Schenk, Mitic, Gahan, & Hengge, 2007; Mitić, Hadler, Gahan, Hengge, & Schenk, 2010; Mitić, Noble, Gahan, Hanson, & Schenk, 2009; Schenk, Peralta, et al., 2008; Smith et al., 2007). Specifically, it could be shown that both pig and red kidney bean PAP hydrolyze both ester bonds in the diester substrate methyl-pNPP in a processive manner (Cox et al., 2007). This observation was initially interpreted in terms of an initial monodentate coordination of the substrate to the divalent metal ion, followed by a nucleophilic attack by a terminal Fe(III)-bound hydroxide. Subsequently, the bridging hydroxide initiates the cleavage of the second ester bond of the substrate in a manner similar to that described in the preceding paragraph (Cox et al., 2007). It has emerged that PAPs may not utilize a terminal, Fe(III)-bound hydroxide but instead employ a water molecule in the second coordination sphere as nucleophile (Merkx & Averill, 1999; Mitić et al., 2010), but the overall mechanistic scheme is not affected by this observation.

The regeneration of the resting form of the enzyme, which requires the removal of the phosphate from the active site, is still not well understood. A plausible sequence involves the rearrangement of the bound phosphate group from tripodal (Fig. 10D) to μ-1,3 coordination (Fig. 10E) via a rotation around the axis formed by the two oxygen atoms of phosphate that are terminally coordinated to the two metal ions. No experimental data for the subsequent steps in regeneration are yet available but a possible mechanism is depicted in Fig. 10F–H. The exchange of the phosphate oxygen atom bound to the divalent metal ion by water leads to a monodentate, Fe(III)-bound phosphate. The subsequent release of the phosphate group enables the M1 site to regain resting state, trigonal-bipyramidal geometry (Fig. 10H).

PAP is, arguably, the most characterized binuclear metallohydrolases to date, a reflection of its direct involvement in bone metabolism (Oddie et al., 2000). Overabundance of PAP in blood serum is a prominent histochemical marker for the diagnosis of osteoporosis, and in transgenic studies with mice it could be demonstrated that the levels of PAP expression are closely linked to bone metabolism. Although the precise *in vivo* substrate(s) for PAP is not yet known, the enzyme's abundance in bone-resorbing osteoclasts have rendered it as an attractive target for novel chemotherapeutics to treat osteoporosis (McGeary, Schenk, & Guddat, 2014). Much of the current research attention thus focuses on the development of PAP inhibitors, an endeavor that is strongly guided by the structural and functional insight described in the preceding paragraphs (McGeary et al., 2014; Vella, McGeary, Gahan, & Schenk, 2010).

6. CONCLUSIONS AND OUTLOOK: AGMATINASE, AN EMERGING TARGET FOR BIOTECHNOLOGICAL APPLICATIONS

Hydrolytic reactions may appear deceivingly simple at first glance. However, as illustrated in the few examples described in this chapter, the mechanisms employed by binuclear metallohydrolases to catalyze hydrolytic bond breakages of ester and amide bonds can be extraordinarily complex. It is this complexity that leads not only to the observed catalytic efficiencies and specificities but also to the functional versatility inherent to this large family of enzymes. While some of these enzymes (e.g., PAPs) have an absolute requirement for two bound metal ions in their active sites to maintain catalytic activity, others (e.g., some MBLs and aminopeptidases) can operate in both mono- and binuclear form, although the bimetallic form is generally more reactive (except in B2-type MBLs where binding of the second metal ion leads to inhibition of the enzyme). Different yet again is a third group of enzymes (exemplified by GpdQ) that form a catalytically active binuclear center only in the presence of a substrate ("substrate-promoted" activation).

Similar to their mechanistic versatility binuclear metallohydrolases also display considerable diversity with respect to the metal ions they prefer or require for catalysis. Some of these enzymes have a very strict requirement for a particular metal ion. An example is the Ni(II)-dependent urease (Benini & Ciurli, 2013; Ciurli, 2007). Others are somewhat less stringent, for instance PAPs, which generally have an Fe(III) in one of their metal ion binding sites, but which are more flexible in the other site (Schenk et al., 2013). Similarly, MBLs generally require Zn(II) *in vivo*, but *in vitro* activity can be reconstituted with a range of transition metal ions, suggesting that the preference for Zn(II) may be predominantly a reflection of bioavailability (Crowder et al., 2006; Page & Badarau, 2008; Phelan et al., 2014). Others yet again do not seem to display any particular preference as *in vitro* activity can be reconstituted with a large number of metal ions; organophosphate-degrading enzymes such as GpdQ exemplify this behavior (Schenk et al., 2012).

The mechanistic diversity of binuclear metallohydrolases is also reflected in the multitude of biological roles they are involved in or implicated with. Consequently, it comes a no surprise that a growing number of these enzymes finds application in biotechnology, as targets for novel chemotherapeutic agents (e.g., PAPs, MBLs, MetAP) or agents for bioremediation (e.g., GpdQ). Novel targets for such applications emerge continuously among this family of enzymes, firmly establishing their position as biotechnologically and biomedically relevant systems, a claim that is both exemplified and substantiated by the agmatinase, an enzyme involved in the biosynthesis of polyamines such as spermidine and spermine, which are essential for compounds for the proliferation, differentiation, and migration of mammalian cells (Agostinelli et al., 2010). Agmatinase catalyzes the hydrolysis of agmatine to putrescine (the precursor for spermidine and spermine). Hence, the enzyme plays a crucial role in regulating the intracellular concentrations of agmatine, a neurotransmitter associated with anticonvulsant-, antineurotoxic-, and antidepressant-like actions (Piletz et al., 2013). To date, only agmatinases from prokaryotic sources have been studied. *E. coli* agmatinase exhibits a strict requirement of Mn(II) for catalytic activity (Carvajal et al., 1999; Salas, Lopez, Uribe, & Carvajal, 2004). Although the crystal structure of *E. coli* agmatinase has not yet been determined, it is believed that its active site contains a binuclear di-Mn(II) center similar to that of arginase and other members of the ureahydrolase group of binuclear metallohydrolases (Cama et al., 2004), a hypothesis that is supported by the crystals structures of the putative agmatinases from

Clostridium difficile and *Burkholderia thailandensis* (Baugh et al., 2013) and the agmatinase from *Deinococcus radiodurans* (Ahn et al., 2004). The only exception to date is the agmatinase from the archaea *Methanocaldococcus jannaschii*, which requires Fe(II) instead of Mn(II) for maximum catalytic activity (Miller, Xu, & White, 2012).

Although no mammalian agmatinase has yet been studied an agmatinase-like protein (ALP) has been detected in rat brain tissue (Uribe, Salas, Enriquez, Orellana, & Carvajal, 2007). ALP has significant agmatinase activity although its amino acid sequence greatly differs from all known ureahydrolases; none of the characteristic Mn(II)-binding residues appear conserved. Nonetheless, following the dialysis against the chelator EDTA ALP activity could be fully reconstituted by incubation with 2 mM Mn(II) at 60 °C (Cofre et al., 2014). The same protocol was previously applied to reactivate the functionally related di-Mn(II) arginases from human and rat (Alarcon et al., 2006; Cama et al., 2004; Carvajal, Torres, Uribe, & Salas, 1995); however, it remains to be shown if ALP is a mono- or binuclear metallohydrolase in its active form. Interestingly, the metal ion requirement of ALP appears more complex than that of other binuclear metallohydrolases. Its C-terminal region (residues 459–510) contains a sequence characteristic for a LIM-domain, a feature that folds into two Zn(II)-stabilized zinc fingers (Feuerstein, Wang, Song, Cooke, & Liebhaber, 1994). Removing the LIM-domain leads to a mutant form of ALP with a 10-fold higher k_{cat} and a three-fold decreased K_m value for agmatine (Castro et al., 2011). Furthermore, the isolated LIM-domain significantly inhibits the agmatinase activity of the truncated ALP mutant, suggesting an autoinhibitory role for this zinc finger moiety. Although the precise interactions between ALP and its LIM-domain are not yet understood, and their functional significance still needs to be evaluated, they widen the mechanistic spectrum employed by binuclear metallohydrolases, and invite future studies that may allow us to fully exploit the tremendous potential, academic, and applied, of this exciting family of enzymes.

ACKNOWLEDGMENTS

The authors like to thank the National Health and Medical Research Council (NH&MRC) and Australian Research Council (ARC) for funding. G. S. is grateful for the award of an ARC Future Fellowship and N. M. would like to thank the Science Foundation Ireland (SFI) for financial support in form of an SFI President of Ireland Young Researcher Award.

REFERENCES

Abriata, L. A., Gonzalez, L. J., & Llarrull, L. I. (2008). Engineered mononuclear variants in *Bacillus cereus* metallo-β-lactamase BcII are inactive. *Biochemistry, 47,* 8590–8599.

Agostinelli, E., Marques, M. P., Calheiros, R., Gil, F. P., Tempera, G., Viceconte, N., et al. (2010). Polyamines: Fundamental characters in chemistry and biology. *Amino Acids, 38,* 393–403.

Ahn, H. J., Kim, K. H., Lee, J., Ha, J. Y., Lee, H. H., Kim, D., et al. (2004). Crystal structure of agmatinase reveals structural conservation and inhibition mechanism of the ureohydrolase superfamily. *The Journal of Biological Chemistry, 279,* 50505–50513.

Alarcon, R., Orellana, M. S., Neira, B., Uribe, E., Garcia, J. R., & Carvajal, N. (2006). Mutational analysis of substrate recognition by human arginase type I—Agmatinase activity of the N130D variant. *The FEBS Journal, 273,* 5625–5631.

Allen, S. H., Nuttleman, P. R., Ketcham, C. M., & Roberts, R. M. (1989). Purification and characterization of human bone tartrate-resistant acid phosphatase. *Journal of Bone and Mineral Research, 4,* 47–55.

Antonyuk, S. V., Olczak, M., Olczak, T., Ciuraszkiewicz, J., & Strange, R. W. (2014). The structure of a purple acid phosphatase involved in plant growth and pathogen defence exhibits a novel immunoglobulin-like fold. *The International Union Crystallography Journal, 1,* 101–109.

Badarau, A., & Page, M. I. (2006). The variation of catalytic efficiency of *Bacillus cereus* metallo-β-lactamase with different active site metal ions. *Biochemistry, 45,* 10654–10666.

Baugh, L., Gallagher, L. A., Patrapuvich, R., Clifton, M. C., Gardberg, A. S., Edwards, T. E., et al. (2013). Combining functional and structural genomics to sample the essential *Burkholderia* structome. *PLoS One, 8,* e53851.

Bebrone, C. (2007). Metallo-β-lactamases (classification, activity, genetic organization, structure, zinc coordination) and their superfamily. *Biochemical Pharmacology, 74,* 1686–1701.

Bebrone, C., Delbrück, H., Kupper, M. B., Schlömer, P., Willmann, C., Frère, J.-M., et al. (2009). The structure of the dizinc subclass B2 metallo-β-lactamase CphA reveals that the second inhibitory zinc ion binds in the histidine site. *Antimicrobial Agents and Chemotherapy, 53,* 4464–4471.

Bebrone, C., Frère, J.-M., Galleni, M., Anne, C., Kerff, F., Garau, G., et al. (2008). Mutational analysis of the zinc- and substrate-binding sites in the CphA metallo-β-lactamase from *Aeromonas hydrophila*. *The Biochemical Journal, 414,* 151–159.

Beck, J. L., McConachie, L. A., Summors, A. C., Arnold, W. N., de Jersey, J., & Zerner, B. (1986). Properties of a purple phosphatase from red kidney bean: A zinc-iron metalloenzyme. *Biochimica et Biophysica Acta, 869,* 61–68.

Ben-Bassat, A., Bauer, K., Chang, S. Y., Myambo, K., Boosman, A., & Chang, S. (1987). Processing of the initiation methionine from proteins: Properties of the *Escherichia coli* methionine aminopeptidase and its gene structure. *Journal of Bacteriology, 169,* 751–757.

Benini, S. F. M., & Ciurli, S. (2013). Urease: *Encyclopedia of metallo proteins*. (pp. 2287-2292) New York: Springer Editions.

Bounaga, S., Laws, A. P., Galleni, M., & Page, M. I. (1998). The mechanism of catalysis and the inhibition of the *Bacillus cereus* zinc-dependent β-lactamase. *The Biochemical Journal, 331,* 703–711.

Bozzo, G. G., Dunn, E. L., & Plaxton, W. C. (2006). Differential synthesis of phosphate-starvation inducible purple acid phosphatase isozymes in tomato (*Lycopersicon esculentum*) suspension cells and seedlings. *Plant, Cell & Environment, 29,* 303–313.

Bozzo, G. G., Raghothama, K. G., & Plaxton, W. C. (2002). Purification and characterization of two secreted purple acid phosphatase isozymes from phosphate-starved tomato (*Lycopersicon esculentum*) cell cultures. *European Journal of Biochemistry, 269,* 6278–6286.

Bozzo, G. G., Raghothama, K. G., & Plaxton, W. C. (2004). Structural and kinetic properties of a novel purple acid phosphatase from phosphate-starved tomato (*Lycopersicon esculentum*) cell cultures. *The Biochemical Journal, 377*, 419–428.
Bradshaw, R. A., & Yi, E. (2002). Methionine aminopeptidases and angiogenesis. *Essays in Biochemistry, 38*, 65–78.
Cama, E., Pethe, S., Boucher, J.-L., Han, S., Emig, F. A., Ash, D. E., et al. (2004). Inhibitor coordination interactions in the binuclear manganese cluster of arginase. *Biochemistry, 43*, 8987–8999.
Campbell, H. D., Dionysius, D. A., Keough, D. T., Wilson, B. E., de Jersey, J., & Zerner, B. (1978). Iron-containing acid phosphatases: Comparison of the enzymes from beef spleen and pig allantoic fluid. *Biochemical and Biophysical Research Communications, 82*, 615–620.
Carfi, A., Pares, S., Duee, E., Galleni, M., Duez, C., Frere, J. M., et al. (1995). The 3-D structure of a zinc metallo-β-lactamase from *Bacillus cereus* reveals a new type of protein fold. *The EMBO Journal, 14*, 4914–4921.
Carvajal, N., López, V., Salas, M., Uribe, E., Herrera, P., & Cerpa, J. (1999). Manganese is essential for catalytic activity of *Escherichia coli* agmatinase. *Biochemical and Biophysical Research Communications, 258*, 808.
Carvajal, N., Torres, C., Uribe, E., & Salas, M. (1995). Interaction of arginase with metal ions: Studies of the enzyme from human liver and comparison with other arginases. *Physiology of Comparative Biochemistry, 112*, 153–159.
Castro, V., Fuentealba, P., Henriquez, A., Vallejos, A., Benitez, J., Lobos, M., et al. (2011). Evidence for an inhibitory LIM domain in a rat brain agmatinase-like protein. *Archives of Biochemistry and Biophysics, 512*, 107–110.
Chang, Y. H., Teichert, U., & Smith, J. A. (1992). Molecular cloning, sequencing, deletion, and overexpression of a methionine aminopeptidase gene from *Saccharomyces cerevisiae*. *The Journal of Biological Chemistry, 267*, 8007–8011.
Chiu, C. H., Lee, C. Z., Lin, K. S., Tam, M. F., & Lin, L. Y. (1999). Amino acid residues involved in the functional integrity of *Escherichia coli* methionine aminopeptidase. *Journal of Bacteriology, 181*, 4686–4689.
Ciurli, S. (2007). Urease: Recent insights on the role of nickel. *Nickel and its surprising impact in nature*. (pp. 241-277) Chichester, UK: John Wiley & Sons, Ltd.
Cofre, J., Montes, P., Vallejos, A., Benitez, J., Garcia, D., Martinez-Oyanedel, J., et al. (2014). Further insight into the inhibitory action of a LIM/double zinc-finger motif of an agmatinase-like protein. *Journal of Inorganic Biochemistry, 132*, 92–95.
Copik, A. J., Waterson, S., Swierczek, S. I., Bennett, B., & Holz, R. C. (2005). Both nucleophile and substrate bind to the catalytic Fe(II)-center in the type-II methionyl aminopeptidase from *Pyrococcus furiosus*. *Inorganic Chemistry, 44*, 1160–1162.
Cox, R. S., Schenk, G., Mitic, N., Gahan, L. R., & Hengge, A. C. (2007). Diesterase activity and substrate binding in purple acid phosphatases. *Journal of the American Chemical Society, 129*, 9550–9551.
Cricco, J. A., Orellano, E. G., Rasia, R. M., Ceccarelli, E. A., & Vila, A. J. (1999). Metallo-β-lactamases: Does it take two to tango? *Coordination Chemistry Reviews, 192*, 519–535.
Crowder, M. W., Spencer, J., & Vila, A. J. (2006). Metallo-β-lactamases: Novel weaponry for antibiotic resistance in bacteria. *Accounts of Chemical Research, 39*, 721–728.
Crowder, M. W., Walsh, T. R., Banovic, L., Pettit, M., & Spencer, J. (1998). Overexpression, purification, and characterization of the cloned metallo-β-lactamase L1 from *Stenotrophomonas maltophilia*. *Antimicrobial Agents and Chemotherapy, 42*, 921–926.
Crowder, M. W., Wang, Z., Franklin, S. L., Zovinka, E. P., & Benkovic, S. J. (1996). Characterization of the metal-binding sites of the β-lactamase from *Bacteroides fragilis*. *Biochemistry, 35*, 12126–12132.
Dal Peraro, M., Llarrull, L. I., & Rothlisberger, U. (2004). Water-assisted reaction mechanism of monozinc β-lactamases. *Journal of the American Chemical Society, 126*, 12661–12668.

Dal Peraro, M., Vila, A. J., & Carloni, P. (2004). Substrate binding to mononuclear metallo-β-lactamase from *Bacillus cereus*. *Proteins*, *54*, 412–423.
Dal Peraro, M., Vila, A. J., Carloni, P., & Klein, M. L. (2007). Role of zinc content on the catalytic efficiency of B1 metallo β-lactamases. *Journal of the American Chemical Society*, *129*, 2808–2816.
Daumann, L. J., Comba, P., Larrabee, J. A., Schenk, G., Stranger, R., Cavigliasso, G., et al. (2013). Synthesis, magnetic properties, and phosphoesterase activity of dinuclear cobalt(II) complexes. *Inorganic Chemistry*, *52*, 2029–2043.
Daumann, L. J., Larrabee, J. A., Ollis, D. L., Schenk, G., & Gahan, L. R. (2014). Immobilization of the enzyme GpdQ on magnetite nanoparticles for organophosphate pesticide bioremediation. *Journal of Inorganic Biochemistry*, *131*, 1–7.
Daumann, L. J., McCarthy, B. Y., Hadler, K. S., Murray, T. P., Gahan, L. R., Larrabee, J. A., et al. (2013). Promiscuity comes at a price: Catalytic versatility vs efficiency in different metal ion derivatives of the potential bioremediator GpdQ. *Biochimica et Biophysica Acta*, *1834*, 425–432.
Daumann, L. J., Schenk, G., & Gahan, L. R. (2014). Metallo-β-lactamases and their biomimetic complexes. *European Journal of Inorganic Chemistry*, 2869–2885.
Davis, J. C., Lin, S. S., & Averill, B. A. (1981). Kinetics and optical spectroscopic studies on the purple acid phosphatase from beef spleen. *Biochemistry*, *20*, 4062–4067.
Day, E. P., David, S. S., Peterson, J., Dunham, W. R., Bonvoisin, J. J., Sands, R. H., et al. (1988). Magnetization and electron paramagnetic resonance studies of reduced uteroferrin and its "EPR-silent" phosphate complex. *The Journal of Biological Chemistry*, *263*, 15561–15567.
de Seny, D., Heinz, U., Wommer, S., Kiefer, M., Meyer-Klaucke, W., Galleni, M., et al. (2001). Metal ion binding and coordination geometry for wild type and mutants of metallo-β-lactamase from *Bacillus cereus* 569/H/9 (BcII): A combined thermodynamic, kinetic, and spectroscopic approach. *The Journal of Biological Chemistry*, *276*, 45065–45078.
del Pozo, J. C., Allona, I., Rubio, V., Leyva, A., de la Pena, A., Aragoncillo, C., et al. (1999). A type 5 acid phosphatase gene from *Arabidopsis thaliana* is induced by phosphate starvation and by some other types of phosphate mobilising/oxidative stress conditions. *The Plant Journal*, *19*, 579–589.
Dismukes, C. G. (1996). Manganese enzymes with binuclear active sites. *Chemical Reviews*, *96*, 2909–2926.
Douangamath, A., Schulz, H., MacSweeney, A., Thormann, M., Treml, A., Pierau, S., et al. (2004). Crystal structures of *Staphylococcus aureus* methionine aminopeptidase complexed with keto heterocycle and aminoketone inhibitors reveal the formation of a tetrahedral intermediate. *Journal of Medicinal Chemistry*, *47*, 1325–1328.
D'Souza, V. M., Bennett, B., Copik, A. J., & Holz, R. C. (2000). Divalent metal binding properties of the methionyl aminopeptidase from *Escherichia coli*. *Biochemistry*, *39*, 3817–3826.
D'Souza, V. M., & Holz, R. C. (1999). The methionyl aminopeptidase from *Escherichia coli* can function as an Fe(II) enzyme. *Biochemistry*, *38*, 11079–11085.
D'Souza, V. M., Swierczek, S. I., Cosper, N. J., Meng, L., Ruebush, S., Copik, A. J., et al. (2002). Kinetic and structural characterization of manganese(II)-loaded methionyl aminopeptidases. *Biochemistry*, *41*, 13096–13105.
Durmus, A., Eicken, C., Spener, F., & Krebs, B. (1999). Cloning and comparative protein modeling of two purple acid phosphatase isozymes from sweet potatoes (*Ipomoea batatas*). *Biochimica et Biophysica Acta*, *1434*, 202–209.
Ek-Rylander, B., Bill, P., Norgard, M., Nilsson, S., & Andersson, G. (1991). Cloning, sequence, and developmental expression of a type 5, tartrate-resistant, acid phosphatase of rat bone. *The Journal of Biological Chemistry*, *266*, 24684–24689.

Ely, F., Foo, J. L., Jackson, C. J., Gahan, L. R., Ollis, D. L., & Schenk, G. (2007). Enzymatic bioremediation: Organophosphate degradation by binuclear metallo-hydrolases. *Current Topics in Biochemical Research, 9*, 63–78.

Estiu, G., Suarez, D., & Merz, K. M. (2006). Quantum mechanical and molecular dynamics simulations of ureases and Zn β-lactamases. *Journal of Computational Chemistry, 27*, 1240–1262.

Evdokimov, A. G., Fairweather, N., Diven, C., Rastogi, V., Grinius, L., Klanke, C., et al. (2007). Serendipitous discovery of novel bacterial methionine aminopeptidase inhibitors. *Proteins, 66*, 538–546.

Fabiane, S. M., Sohi, M. K., Wan, T., Payne, D. J., Bateson, J. H., Mitchell, T., et al. (1998). Crystal structure of the zinc-dependent β-lactamase from *Bacillus cereus* at 1.9 Å resolution: Binuclear active site with features of a mononuclear enzyme. *Biochemistry, 37*, 12404–12411.

Fast, W., Wang, Z., & Benkovic, S. J. (2001). Familial mutations and zinc stoichiometry determine the rate-limiting step of nitrocefin hydrolysis by metallo-β-lactamase from *Bacteroides fragilis*. *Biochemistry, 40*, 1640–1650.

Feuerstein, R., Wang, X., Song, D., Cooke, N. E., & Liebhaber, S. A. (1994). The LIM/double zinc-finger motif functions as a protein dimerization domain. *Proceedings of the National Academy of Sciences of the United States of America, 91*, 10655–10659.

Flanagan, J. U., Cassady, A. I., Schenk, G., Guddat, L. W., & Hume, D. A. (2006). Identification and molecular modeling of a novel, plant-like, human purple acid phosphatase. *Gene, 377*, 12–20.

Folkman, J. (1995). Angiogenesis in cancer, vascular, rheumatoid and other disease. *Nature Medicine, 1*, 27–30.

Frere, J. M., Galleni, M., Bush, K., & Dideberg, O. (2005). Is it necessary to change the classification of β-lactamases? *The Journal of Antimicrobial Chemotherapy, 55*, 1051–1053.

Funhoff, E. G., Klaassen, C. H., Samyn, B., Van Beeumen, J., & Averill, B. A. (2001). The highly exposed loop region in mammalian purple acid phosphatase controls the catalytic activity. *ChemBioChem, 2*, 355–363.

Garau, G., Bebrone, C., Anne, C., Galleni, M., Frère, J.-M., & Dideberg, O. (2005). A metallo-β-lactamase enzyme in action: Crystal structures of the monozinc carbapenemase CphA and its complex with biapenem. *Journal of Molecular Biology, 345*, 785–795.

Garau, G., García-Sáez, I., Bebrone, C., Anne, C., Mercuri, P., Galleni, M., et al. (2004). Update of the standard numbering scheme for class B β-lactamases. *Antimicrobial Agents and Chemotherapy, 48*, 2347–2349.

Ghanem, E., Li, Y., Xu, C., & Raushel, F. M. (2007). Characterization of a phosphodiesterase capable of hydrolyzing EA 2192, the most toxic degradation product of the nerve agent VX. *Biochemistry, 46*, 9032–9040.

Ghosh, M., Grunden, A. M., Dunn, D. M., Weiss, R., & Adams, M. W. (1998). Characterization of native and recombinant forms of an unusual cobalt-dependent proline dipeptidase (Prolidase) from the hyperthermophilic *Archaeon Pyrococcus furiosus*. *Journal of Bacteriology, 180*, 4781–4789.

Guddat, L. W., McAlpine, A. S., Hume, D., Hamilton, S., de Jersey, J., & Martin, J. L. (1999). Crystal structure of mammalian purple acid phosphatase. *Structure, 7*, 757–767.

Hadler, K. S., Gahan, L. R., Ollis, D. L., & Schenk, G. (2010). The bioremediator glycerophosphodiesterase employs a non-processive mechanism for hydrolysis. *Journal of Inorganic Biochemistry, 104*, 211–213.

Hadler, K. S., Mitić, N., Ely, F., Hanson, G. R., Gahan, L. R., Larrabee, J. A., et al. (2009). Structural flexibility enhances the reactivity of the bioremediator glycerophosphodiesterase by fine-tuning its mechanism of hydrolysis. *Journal of the American Chemical Society, 131*, 11900–11908.

Hadler, K. S., Mitic, N., Yip, S. H., Gahan, L. R., Ollis, D. L., Schenk, G., et al. (2010). Electronic structure analysis of the dinuclear metal center in the bioremediator glycerophosphodiesterase (GpdQ) from *Enterobacter aerogenes*. *Inorganic Chemistry*, *49*, 2727–2734.

Hadler, K. S., Tanifum, E. A., Yip, S., Mitić, N., Guddat, L. W., Jackson, C. J., et al. (2008). Substrate-promoted formation of a catalytically competent binuclear center and regulation of reactivity in glycerophosphodiesterase from *Enterobacter aerogenes*. *Journal of the American Chemical Society*, *130*, 14129–14138.

Hawk, M. J., Breece, R. M., Hajdin, C. E., Bender, K. M., Zhenxin, H., Costello, A. L., et al. (2009). Differential binding of Co(II) and Zn(II) to metallo-β-lactamase Bla2 from *Bacillus anthracis*. *Journal of the American Chemical Society*, *131*, 10753–10762.

Hayman, A. R., & Cox, T. M. (1994). Purple acid phosphatase of the human macrophage and osteoclast. Characterization, molecular properties, and crystallization of the recombinant di-iron-oxo protein secreted by baculovirus-infected insect cells. *The Journal of Biological Chemistry*, *269*, 1294–1300.

Hayman, A. R., Warburton, M. J., Pringle, J. A., Coles, B., & Chambers, T. J. (1989). Purification and characterization of a tartrate-resistant acid phosphatase from human osteoclastomas. *The Biochemical Journal*, *261*, 601–609.

Heinz, U., & Adolph, H. W. (2004). Metallo-β-lactamases: Two binding sites for one catalytic metal ion? *Cellular and Molecular Life Sciences*, *61*, 2827–2839.

Hemmingsen, L., Damblon, C., Antony, J., Jensen, M., Adolph, H. W., Wommer, S., et al. (2001). Dynamics of mononuclear cadmium β-lactamase revealed by the combination of NMR and PAC spectroscopy. *Journal of the American Chemical Society*, *123*, 10329–10335.

Hernandez-Valladares, M., Felici, A., Weber, G., Adolph, H. W., Zeppezauer, M., Rossolini, G. M., et al. (1997). Zn(II) dependence of the *Aeromonas hydrophila* AE036 metallo-β-lactamase activity and stability. *Biochemistry*, *36*, 11534–11541.

Höpfner, K. P., Karcher, A., Craig, L., Woo, T. T., Carney, J. P., & Tainer, J. A. (2001). Structural biochemistry and interaction architecture of the DNA double-strand break repair Mre11 nuclease and Rad50-ATPase. *Cell*, *105*, 473–485.

Hou, C., Phelan, E., Miraula, M., Ollis, D., Schenk, G., & Mitić, N. (2014). Unusual metallo-β-lactamases may constitute a new subgroup in this family of enzymes. *American Journal of Molecular Biology*, *4*, 11–15.

Hu, X., Addlagatta, A., Matthews, B. W., & Liu, J. O. (2006). Identification of pyridinylpyrimidines as inhibitors of human methionine aminopeptidases. *Angewandte Chemie International Edition*, *45*, 3772–3775.

Huang, M., Xie, S.-X., Ma, Z.-Q., Huang, Q.-Q., Nan, F.-J., & Ye, Q.-Z. (2007). Inhibition of monometalated methionine aminopeptidase: Inhibitor discovery and crystallographic analysis. *Journal of Medicinal Chemistry*, *50*, 5735–5742.

Jackson, C. J., Carr, P. D., Liu, J. W., Watt, S. J., Beck, J. L., & Ollis, D. L. (2007). The structure and function of a novel glycerophosphodiesterase from *Enterobacter aerogenes*. *Journal of Molecular Biology*, *367*, 1047–1062.

Jackson, C. J., Hadler, K. S., Carr, P. D., Oakley, A. J., Yip, S., Schenk, G., et al. (2008). Malonate-bound structure of the glycerophosphodiesterase from *Enterobacter aerogenes* (GpdQ) and characterization of the native Fe(II) metal-ion preference. *Acta Crystallographica*, *F64*, 681–685.

Ketcham, C. M., Baumbach, G. A., Bazer, F. W., & Roberts, R. M. (1985). The type 5, acid phosphatase from spleen of humans with hairy cell leukemia. Purification, properties, immunological characterization, and comparison with porcine uteroferrin. *The Journal of Biological Chemistry*, *260*, 5768–5776.

Ketcham, C. M., Roberts, R. M., Simmen, R. C., & Nick, H. S. (1989). Molecular cloning of the type 5, iron-containing, tartrate-resistant acid phosphatase from human placenta. *The Journal of Biological Chemistry*, *264*, 557–563.

Klabunde, T., Sträter, N., Fröhlich, R., Witzel, H., & Krebs, B. (1996). Mechanism of Fe(III)-Zn(II) purple acid phosphatase based on crystal structures. *Journal of Molecular Biology, 259*, 737–748.

Klinkenberg, M., Ling, C., & Chang, Y. H. (1997). A dominant negative mutation in *Saccharomyces cerevisiae* methionine aminopeptidase-1 affects catalysis and interferes with the function of methionine aminopeptidase-2. *Archives of Biochemistry and Biophysics, 347*, 193.

Knöfel, T., & Sträter, N. (1999). X-ray structure of the *Escherichia coli* periplasmic 5'-nucleotidase containing a dimetal catalytic site. *Nature Structural Biology, 6*, 448–453.

Laraki, N., Franceschini, N., Rossolini, G. M., Santucci, P., Meunier, C., de Pauw, E., et al. (1999). Biochemical characterization of the *Pseudomonas aeruginosa* 101/1477 metallo-β-lactamase IMP-1 produced by *Escherichia coli*. *Antimicrobial Agents and Chemotherapy, 43*, 902–906.

Larrabee, J. A., Alessi, C. M., Asiedu, E. T., Cook, J. O., Hoerning, K. R., Klingler, L. J., et al. (1997). Magnetic circular dichroism spectroscopy as a probe of geometric and electronic structure of cobalt(II)-substituted proteins: Ground-state zero- field splitting as a coordination number indicator. *Journal of the American Chemical Society, 119*, 4182–4196.

Larson, T. J., Ehrmann, M., & Boos, W. (1983). Periplasmic glycerophosphodiester phosphodiesterase of *Escherichia coli*, a new enzyme of the glp regulon. *The Journal of Biological Chemistry, 258*, 5428–5432.

Leopoldini, M., Russo, N., & Toscano, M. (2007). Which one among Zn(II), Co(II), Mn(II), and Fe(II) is the most efficient ion for the methionine aminopeptidase catalyzed reaction? *Journal of the American Chemical Society, 129*, 7776–7784.

Li, J.-Y., Li, J., Chen, L.-L., Cui, Y.-M., Luo, Q.-L., Nan, F.-J., et al. (2003). Specificity for inhibitors of metal-substituted methionine aminopeptidase. *Biochemical and Biophysical Research Communications, 307*, 172–179.

Lindqvist, Y., Johansson, E., Kaija, H., Vihko, P., & Schneider, G. (1999). Three-dimensional structure of a mammalian purple acid phosphatase at 2.2 Ångstrom resolution with a μ-(hydr)oxo bridged di-iron center. *Journal of Molecular Biology, 291*, 135–147.

Ljusberg, J., Ek-Rylander, B., & Andersson, G. (1999). Tartrate-resistant purple acid phosphatase is synthesized as a latent proenzyme and activated by cysteine proteinases. *The Biochemical Journal, 343*, 63–69.

Llarrull, L. I., Tioni, M. F., Kowalski, J., Bennett, B., & Vila, A. J. (2007). Evidence for a dinuclear active site in the metallo-β-lactamase BcII with substoichiometric Co(II): A new model for metal uptake. *The Journal of Biological Chemistry, 282*, 30586–30595.

Lowther, W. T., & Matthews, B. W. (2002). Metalloaminopeptidases: Common functional themes in disparate structural surroundings. *Chemical Reviews, 102*, 4581–4607.

Lowther, W. T., Orville, A. M., Madden, D. T., Lim, S., Rich, D. H., & Matthews, B. W. (1999). *Escherichia coli* methionine aminopeptidase: Implications of crystallographic analyses of the native, mutant, and inhibited enzymes for the mechanism of catalysis. *Biochemistry, 38*, 7678–7688.

Lowther, W. T., Zhang, Y., Sampson, P. B., Honek, J. F., & Matthews, B. W. (1999). Insights into the mechanism of *Escherichia coli* methionine aminopeptidase from the structural analysis of reaction products and phosphorus-based transition-state analogues. *Biochemistry, 38*, 14810–14819.

Marshall, K., Nash, K., Haussman, G., Cassady, I., Hume, D., de Jersey, J., et al. (1997). Recombinant human and mouse purple acid phosphatases: Expression and characterization. *Archives of Biochemistry and Biophysics, 345*, 230–236.

McGeary, R. P., Schenk, G., & Guddat, L. W. (2014). The applications of binuclear metallohydrolases in medicine: Recent advances in the design and development of novel drug leads for purple acid phosphatases, metallo-β-lactamases and arginases. *European Journal of Medicinal Chemistry, 76*, 132–144.

Meng, L., Ruebush, S., D'Souza, V. M., Copik, A. J., Tsunasawa, S., & Holz, R. C. (2002). Overexpression and divalent metal binding properties of the methionyl aminopeptidase from *Pyrococcus furiosus*. *Biochemistry, 41*, 7199–7208.

Mercuri, P. S., Bouillenne, F., Boschi, L., Lamotte-Brasseur, J., Amicosante, G., Devreese, B., et al. (2001). Biochemical characterization of the FEZ-1 metallo-β-lactamase of *Legionella gormanii* ATCC 33297T produced in *Escherichia coli*. *Antimicrobial Agents and Chemotherapy, 45*, 1254–1262.

Merkx, M., & Averill, B. A. (1999). Probing the role of the trivalent metal in phosphate ester hydrolysis. *Journal of the American Chemical Society, 121*, 6683–6689.

Merkx, M., Pinkse, M. W., & Averill, B. A. (1999). Evidence for nonbridged coordination of p-nitrophenyl phosphate to the dinuclear Fe(III)-M(II) center in bovine spleen purple acid phosphatase during enzymatic turnover. *Biochemistry, 38*, 9914–9925.

Meyers, R. A. (1996). *Encyclopedia of molecular biology and molecular medicine*. Weinheim: VCH Publishers.

Miller, D., Xu, H., & White, R. H. (2012). A new subfamily of agmatinases present in methanogenic *Archaea* is Fe(II) dependent. *Biochemistry, 51*, 3067–3078.

Mitić, N., Hadler, K. S., Gahan, L. R., Hengge, A. C., & Schenk, G. (2010). The divalent metal ion in the active site of uteroferrin modulates substrate binding and catalysis. *Journal of the American Chemical Society, 132*, 7049–7054.

Mitić, N., Noble, C. J., Gahan, L. R., Hanson, G. R., & Schenk, G. (2009). Metal ion mutagenesis—Conversion of a purple acid phosphatase from sweet potato to a neutral phosphatase with the formation of an unprecedented catalytically competent Mn^{II}Mn^{II} active site. *Journal of the American Chemical Society, 131*, 8173–8179.

Mitić, N., Smith, S. J., Neves, A., Guddat, L. W., Gahan, L. R., & Schenk, G. (2006). The catalytic mechanisms of binuclear metallohydrolases. *Chemical Reviews, 106*, 3338–3363.

Mitić, N., Valizadeh, M., Leung, E. W. W., de Jersey, J., Hamilton, S., Hume, D. A., et al. (2005). Human tartrate-resistant acid phosphatase becomes an effective ATPase upon proteolytic activation. *Archives of Biochemistry and Biophysics, 439*, 154–164.

Morán-Barrio, J., Viale, A. M., Vila, A. J., González, J. M., Lisa, M. N., Costello, A. L., et al. (2007). The metallo-β-lactamase GOB is a mono-Zn(II) enzyme with a novel active site. *The Journal of Biological Chemistry, 282*, 18286–18293.

Nakazato, H., Okamoto, T., Nishikoori, M., Washio, K., Morita, N., Haraguchi, K., et al. (1998). The glycosylphosphatidylinositol-anchored phosphatase from *Spirodela oligorrhiza* is a purple acid phosphatase. *Plant Physiology, 118*, 1015–1020.

Nishikoori, M., Washio, K., Hase, A., Morita, N., & Okuyama, H. (2001). Cloning and characterization of cDNA of the GPI-anchored purple acid phosphatase and its root tissue distribution in *Spirodela oligorrhiza*. *Physiologia Plantarum, 113*, 241–248.

Oddie, G. W., Schenk, G., Angel, N. Z., Walsh, N., Guddat, L. W., de Jersey, J., et al. (2000). Structure, function and regulation of tartrate-resistant acid phosphatase. *Bone, 27*, 575–584.

Oefner, C., Thormann, M., Wadman, S., Dale, G. E., Douangamath, A., D'Arcy, A., et al. (2003). The 1.15 Å crystal structure of the *Staphylococcus aureus* methionyl-aminopeptidase and complexes with triazole based inhibitors. *Journal of Molecular Biology, 332*, 13–21.

Outten, C. E., & O'Halloran, T. V. (2001). Femtomolar sensitivity of metalloregulatory proteins controlling zinc homeostasis. *Science, 292*, 2488–2492.

Page, M. I., & Badarau, A. (2008). The mechanisms of catalysis by metallo β-lactamases. *Bioinorganic Chemistry and Applications*, 1–14.

Paul-Soto, R., Hernandez-Valladares, M., Galleni, M., Bauer, R., Zeppezauer, M., Frere, J. M., et al. (1998). Mono- and binuclear Zn-β-lactamase from *Bacteroides fragilis*: Catalytic and structural roles of the zinc ions. *FEBS Letters, 438*, 137–140.

Paul-Soto, R., Zeppezauer, M., Adolph, H.-W., Bauer, R., Frère, J.-M., Galleni, M., et al. (1999). Mono- and binuclear Zn(II)-β-lactamase. Role of the conserved cysteine in the catalytic mechanism. *The Journal of Biological Chemistry, 274*, 13242–13249.

Pedroso, M. M., Ely, F., Lonhienne, T., Gahan, L. R., Ollis, D. L., Guddat, L. W., et al. (2014). Determination of the catalytic activity of binuclear metallohydrolases using isothermal titration calorimetry. *Journal of Biological Inorganic Chemistry, 19*, 389–398.

Phelan, E. K., Miraula, M., Selleck, C., Ollis, D. L., Schenk, G., & Mitić, N. (2014). Metallo-β-lactamases: A major threat to human health. *American Journal of Molecular Biology, 4*, 89–104. http://dx.doi.org/10.4236/ajmb.2014.43011.

Piletz, J. E., Li, J., Liu, P., Molderings, G. J., Rodrigues, A. L. S., Satriano, J., et al. (2013). Agmatine: Clinical applications after 100 years in translation. *Drug Discovery Today, 18*, 880–893.

Raetz, C. R. (1986). Molecular genetics of membrane phospholipid synthesis. *Annual Review of Genetics, 20*, 253–295.

Saavedra, M. J., Peixe, L., Sousa, J. C., Henriques, I., Alves, A., & Correia, A. (2003). Sfh-I, a subclass B2 metallo-β-lactamase from a *Serratia fonticola* environmental isolate. *Antimicrobial Agents and Chemotherapy, 47*, 2330–2333.

Salas, M., Lopez, V., Uribe, E., & Carvajal, N. (2004). Studies on the interaction of *Escherichia coli* agmatinase with manganese ions: Structural and kinetic studies of the H126N and H151N variants. *Journal of Inorganic Biochemistry, 98*, 1032–1036.

Schenk, G., Boutchard, C. L., Carrington, L. E., Noble, C. J., Moubaraki, B., Murray, K. S., et al. (2001). A purple acid phosphatase from sweet potato contains an antiferromagnetically coupled binuclear Fe-Mn center. *The Journal of Biological Chemistry, 276*, 19084–19088.

Schenk, G., Carrington, L. E., Hamilton, S. E., de Jersey, J., & Guddat, L. W. (1999). Crystallization and preliminary X-ray diffraction data for a purple acid phosphatases from sweet potato. *Acta Crystallographica, D55*, 2051–2052.

Schenk, G., Elliott, T. W., Leung, E., Carrington, L. E., Mitić, N., Gahan, L. R., et al. (2008). Crystal structures of a purple acid phosphatase, representing different steps of this enzyme's catalytic cycle. *BMC Structural Biology, 8*, 6.

Schenk, G., Gahan, L. R., Carrington, L. E., Mitic, N., Valizadeh, M., Hamilton, S. E., et al. (2005). Phosphate forms an unusual tripodal complex with the Fe-Mn center of sweet potato purple acid phosphatase. *Proceedings of the National Academy of Sciences of the United States of America, 102*, 273–278.

Schenk, G., Ge, Y., Carrington, L. E., Wynne, C. J., Searle, I. R., Carroll, B. J., et al. (1999). Binuclear metal centers in plant purple acid phosphatases: Fe-Mn in sweet potato and Fe-Zn in soybean. *Archives of Biochemistry and Biophysics, 370*, 183–189.

Schenk, G., Guddat, L. W., Ge, Y., Carrington, L. E., Hume, D. A., Hamilton, S., et al. (2000). Identification of mammalian-like purple acid phosphatases in a wide range of plants. *Gene, 250*, 117–125.

Schenk, G., Korsinczky, M., Hume, D., Hamilton, S., & de Jersey, J. (2000). Purple acid phosphatases from bacteria: Similarities to mammalian and plant enzymes. *Gene, 255*, 419–424.

Schenk, G., Mitic, N., Gahan, L. R., Ollis, D. L., McGeary, R. P., & Guddat, L. W. (2012). Binuclear metallohydrolases: Complex mechanistic strategies for a simple chemical reaction. *Accounts of Chemical Research, 45*, 1593–1603.

Schenk, G., Mitic, N., Hanson, G. R., & Comba, P. (2013). Purple acid phosphatase: A journey into the function and mechanism of a colorful enzyme. *Coordination Chemistry Reviews, 257*, 473–482.

Schenk, G., Peralta, R. A., Batista, S. C., Bortoluzzi, A. J., Szpoganicz, B., Dick, A., et al. (2008). Probing the role of the divalent metal ion in uteroferrin using metal ion replacement and a comparison to isostructural biomimetics. *Journal of Biological Inorganic Chemistry, 13*, 139–155.

Schiffmann, R., Heine, A., Klebe, G., & Klein, C. D. P. (2005). Metal ions as cofactors for the binding of inhibitors to methionine aminopeptidase: A critical view of the relevance of in vitro metalloenzyme assays. *Angewandte Chemie International Edition, 44*, 3620–3623.

Sharma, N. P., Hajdin, C., Chandrasekar, S., Bennett, B., Yang, K.-W., & Crowder, M. W. (2006). Mechanistic studies on the mononuclear Zn(II)-containing metallo-β-lactamase ImiS from *Aeromonas sobria*. *Biochemistry, 45*, 10729–10738.

Smith, S. J., Casellato, A., Hadler, K. S., Mitić, N., Riley, M. J., Bortoluzzi, A. J., et al. (2007). The reaction mechanism of binuclear metallohydrolases: The catalytic role of the metal ions. *Journal of Biological Inorganic Chemistry, 12*, 1207–1220.

Smoukov, S. K., Quaroni, L., Wang, X., Doan, P. E., Hoffman, B. M., & Que, L., Jr. (2002). Electro-nuclear double resonance spectroscopic evidence for a hydroxo-bridge nucleophile involved in catalysis by a dinuclear hydrolase. *Journal of the American Chemical Society, 124*, 2595–2603.

Spencer, J., Clarke, A. R., & Walsh, T. R. (2001). Novel Mechanism of hydrolysis of therapeutic β-lactams by *Stenotrophomonas maltophilia* L1 metallo-β-lactamase. *The Journal of Biological Chemistry, 276*, 33638–33644.

Stoczko, M., Frère, J.-M., Rossolini, G. M., & Docquier, J.-D. (2006). Postgenomic scan of metallo-β-lactamase homologues in *Rhizobacteria*: Identification and characterization of BJP-1, a subclass B3 ortholog from *Bradyrhizobium japonicum*. *Antimicrobial Agents and Chemotherapy, 50*, 1973–1981.

Sträter, N., Beate, J., Scholte, M., Krebs, B., Duff, A. P., Langley, D. B., et al. (2005). Crystal structures of recombinant human purple acid phosphatase with and without an inhibitory conformation of the repression loop. *Journal of Molecular Biology, 351*, 233–246.

Sträter, N., Klabunde, T., Tucker, P., Witzel, H., & Krebs, B. (1995). Crystal structure of a purple acid phosphatase containing a dinuclear Fe(III)-Zn(II) active site. *Science, 268*, 1489–1492.

Sträter, N., Lipscomb, W. N., Klabunde, T., & Krebs, B. (1996). Two-metal ion catalysis in enzymatic acyl- and phosphoryl-transfer reactions. *Angewandte Chemie, 35*, 2024–2055.

Twitchett, M. B., Schenk, G., Aquino, M. A., Yiu, D. T., Lau, T. C., & Sykes, A. G. (2002). Reactivity of M(II) metal-substituted derivatives of pig purple acid phosphatase (uteroferrin) with phosphate. *Inorganic Chemistry, 41*, 5787–5794.

Ullah, J. H., Walsh, T. R., Taylor, I. A., Emery, D. C., Verma, C. S., Gamblin, S. J., et al. (1998). The crystal structure of the L1 metallo-beta-lactamase from *Stenotrophomonas maltophilia* at 1.7 Å resolution. *Journal of Molecular Biology, 284*, 125.

Uribe, E., Salas, M., Enriquez, S., Orellana, M. S., & Carvajal, N. (2007). Cloning and functional expression of a rodent brain cDNA encoding a novel protein with agmatinase activity, but not belonging to the arginase family. *Archives of Biochemistry and Biophysics, 461*, 146–150.

Veljanovski, V., Vanderbeld, B., Knowles, V. L., Snedden, W. A., & Plaxton, W. C. (2006). Biochemical and molecular characterization of AtPAP26, a vacuolar purple acid phosphatase up-regulated in phosphate-deprived *Arabidopsis* suspension cells and seedlings. *Plant Physiology, 142*, 1282–1293.

Vella, P., McGeary, R. P., Gahan, L. R., & Schenk, G. (2010). Tartrate-resistant acid phosphatases, a target for novel anti-osteoporotic chemotherapeutics. *Current Enzyme Inhibition, 6*, 118–129.

Vella, P., Phelan, E., Leung, E. W. W., Ely, F., Ollis, D. L., McGeary, R. P., et al. (2013). Identification and characterization of an unusual metallo-β-lactamase from *Serratia proteamaculans*. *Journal of Biological Inorganic Chemistry, 18*, 855–863.

Walker, K. W., & Bradshaw, R. A. (1998). Yeast methionine aminopeptidase I can utilize either Zn(II) or Co(II) as a cofactor: A case of mistaken identity? *Protein Science, 7*, 2684–2687.

Wang, J., Henkin, J., Sheppard, G. S., Lou, P., Kawai, M., Park, C., et al. (2003). Physiologically relevant metal cofactor for methionine aminopeptidase-2 is manganese. *Biochemistry, 42*, 5035–5042.

Wang, X., & Que, L., Jr. (1998). Extended X-ray absorption fine structure studies of the anion complexes of FeZn uteroferrin. *Biochemistry, 37*, 7813–7821.

Wang, X., Randall, C. R., True, A. E., & Que, L., Jr. (1996). X-ray absorption spectroscopic studies of the FeZn derivative of uteroferrin. *Biochemistry, 35*, 13946–13954.

Wommer, S., Adolph, H.-W., Rival, S., Heinz, U., Galleni, M., Frère, J.-M., et al. (2002). Substrate-activated zinc binding of metallo-β-lactamases: Physiological importance of the mononuclear enzymes. *The Journal of Biological Chemistry, 277*, 24142–24147.

Xie, S. X., Huang, W. J., Ma, Z. Q., Huang, M., Hanzlik, R. P., & Ye, Q. Z. (2006). Structural analysis of metalloform-selective inhibition of methionine aminopeptidase. *Acta Crystallographica, D62*, 425–432.

Yang, Y.-S., McCormick, J. M., & Solomon, E. I. (1997). Circular dichroism and magnetic circular dichroism studies of the mixed-valence binuclear non-heme iron active site in uteroferrin and its anion complexes. *Journal of the American Chemical Society, 119*, 11832–11842.

Ye, Q. Z., Xie, S. X., Huang, M., Huang, W. J., Lu, J. P., & Ma, Z. Q. (2004). Metalloform-selective inhibitors of *Escherichia coli* methionine aminopeptidase and X-ray structure of a Mn(II)-form enzyme complexed with an inhibitor. *Journal of the American Chemical Society, 126*, 13940–13941.

Ye, Q. Z., Xie, S. X., Ma, Z. Q., Huang, M., & Hanzlik, R. P. (2006). Structural basis of catalysis by monometalated methionine aminopeptidase. *Proceedings of the National Academy of Sciences of the United States of America, 103*, 9470–9475.

Zimmermann, P., Regierer, B., Kossmann, J., Frossard, E., Amrhein, N., & Bucher, M. (2004). Differential expression of three purple acid phosphatases from potato. *Plant Biology, 6*, 519–528.

CHAPTER FOUR

Applications of Quantum Mechanical/Molecular Mechanical Methods to the Chemical Insertion Step of DNA and RNA Polymerization

Lalith Perera*,1, William A. Beard*, Lee G. Pedersen†, Samuel H. Wilson*

*Laboratory of Structural Biology, National Institution of Environmental Health Sciences, Research Triangle Park, North Carolina, USA
†Department of Chemistry, University of North Carolina at Chapel Hill, Chapel Hill, North Carolina, USA
1Corresponding author: e-mail address: pereral2@niehs.nih.gov

Contents

1. Introduction — 85
2. Methods for Describing Reactive Pathways — 85
3. DNA Polymerase β — 87
   3.1 Abashkin, Erickson, and Burt (2001): A cluster QM calculation at steps along a proposed reaction path — 87
   3.2 Rittenhouse, Apostoluk, Miller, and Straatsma (2003): A cluster QM calculation on the prechemistry complex of Pol β — 89
   3.3 Radhakrishnan and Schlick (2006): QM/MM study of Pol β (based on PDB = 1BPY; Sawaya et al., 1997) — 89
   3.4 Lin et al. (2006): QM/MM study of Pol β (based on PDB = 2FMS; Batra et al., 2006) — 90
   3.5 Lin et al. (2008): Pol β incorrect insertion QM/MM study (based on PDB = 2C2K; Batra et al., 2008) — 94
   3.6 Batra et al. (2013): A QM/MM study based on the X-ray crystal structure of the Pol β variant D256E — 95
   3.7 The broader picture: Pol β — 97
4. Application of QM/MM to Systems Similar to Pol β — 98
   4.1 Pol λ insertion — 98
   4.2 Dpo4 insertion — 100
   4.3 Pol κ insertion — 101
5. RNA Polymerase — 103
6. Thoughts for Future QM/MM Simulations on Ternary Substrate Complexes of Nucleic Acid Polymerases — 106

Advances in Protein Chemistry and Structural Biology, Volume 97
ISSN 1876-1623
http://dx.doi.org/10.1016/bs.apcsb.2014.10.001

7. Conclusions	109
Acknowledgments	109
References	109

Abstract

We review theoretical attempts to model the chemical insertion reactions of nucleoside triphosphates catalyzed by the nucleic acid polymerases using combined quantum mechanical/molecular mechanical methodology. Due to an existing excellent database of high-resolution X-ray crystal structures, the DNA polymerase β system serves as a useful template for discussion and comparison. The convergence of structures of high-quality complexes and continued developments of theoretical techniques suggest a bright future for understanding the global features of nucleic acid polymerization.

ABBREVIATIONS

Ab initio An invented term, by Robert Parr and coworkers, which means "from the beginning" in quantum mechanics. The sense in which the term is used is to imply "more accurate" and "from first principles" as opposed to empirical or semiempirical.

DFT Density functional theory, an alternate theoretical procedure for determining energies and structures of molecules. The theory is still in a developmental stage.

Fidelity The accuracy of DNA or RNA polymerization and is the reciprocal of the misinsertion frequency.

G0X ($X = 1, 3, 9, \ldots$) Computer packages of Gaussian, Inc., for the calculation of properties of molecules, including *ab initio* calculation of the energy. The programs trace to the vision of John Pople.

Insertion The bonding of the nucleotide fragment (NMP) of an NTP to the O3$'$ growing nucleic acid terminus.

NTP Deoxyribonucleoside or ribonucleoside triphosphate.

PDB Protein Data Bank.

PM3MM A semiempirical quantum mechanical code for calculation of properties of molecules. It contains optional corrections for HCON linkages. It is fitted to thermochemical and spectroscopic data. The programs trace to the vision of M.J.S. Dewar.

PME Particle Mesh Ewald, a fast Fourier transform method to compute the electrostatic energy of a periodic array of charges (see Darden, York, & Pedersen, 1993; Essmann et al., 1995).

PMEMD An efficient, modified version of the Sander module of AMBER to perform PME calculations (written by Bob Duke in the laboratories of Tom Darden and Lee Pedersen).

Pol β A moderate fidelity member of the X-family of DNA polymerases for which many high-resolution crystal structures exist.

QM/MM A procedure that divides a large molecular system into a quantum mechanical part (the smaller part) and a molecular mechanical part (the larger part) with various techniques used to minimize the disruption that occurs at the boundary. The quantum mechanical calculation allows for bond breaking and bond forming.

1. INTRODUCTION

The evolution of modern X-ray crystallography has led to a rapidly increasing wealth of information about the three-dimensional structures of both DNA and RNA polymerases (Wu, Beard, Pedersen, & Wilson, 2014). The resulting high-resolution structures, capture intermediates that span the reactants to products path, have created a fertile ground for computational theoreticians to develop, test, and apply methods that can expose the finer details of the bonds that form and break. In this chapter, we focus on several nucleic acid polymerization systems for which sufficient structures exist to reasonably explore the energy requirements of possible pathways.

An overview of the current state of the rapidly changing knowledge of DNA repair enzymes can be found in the more recent review by Sobol (2014) and the more structural review by Wu et al. (2014). The nucleic acid (DNA and RNA) polymerase chemistry and structures have been systematically discussed in the recent volume edited by Murakami and Trakselis (2014). The basic premise is that good structures of pertinent complexes may serve as starting points for theoretical studies along a reaction path between reactant (prechemistry) and product.

The 2013 Nobel prize for Chemistry recognized the pioneering theoretical developments by M. Karplus, M. Levitt, and A. Warshel, the central contribution being the quantum mechanical/molecular mechanical (QM/MM) approach to break a large biological molecular system up into a QM core where bonds can break and form and an MM system where covalent bonds are not formed or broken. In the latter case, classical force fields can be used to describe the motion and long-range electrostatics (Bash et al., 1991; Warshell & Levitt, 1976). Recent reviews that focus on the development of the method are available (Groenhof, 2013; Lin & Truhlar, 2007; Konig, Hudson, Boresch, & Woodcock, 2014; Senn & Thiel, 2009).

2. METHODS FOR DESCRIBING REACTIVE PATHWAYS

There are many variations of the QM/MM methodology. Most have to do with the manner in which the boundary between the QM and classical mechanical regions is treated. (There is the story of the party of mathematicians that you happen to be attending. One of the groups is expounding on a particular idea. The best way to become included in the group, the story goes, is to ask, at a particular point of emphasis in the exposition: "But, what happens at the boundary?"). How to treat the boundary and what happens at the boundary are the essential questions that we face. Let us consider an example of a popular version of QM/MM—the ONIOM method of the Morokuma group (Vreven et al., 2006). The ONIOM method considers four energies—E(QM,all), E(MM,all), E(QM,Q), and E(MM,Q): the QM and MM energies of the complete system and the QM and MM energies of the specified quantum region, Q, respectively. Then the boundary is chosen (in principle) in such a way that the differences between E(QM,all) and E(MM,all) and between E(QM,Q) and E(MM,Q) are equal or nearly so. Then, we would have

$$E(\text{QM, all}) = E(\text{MM, all}) + [E(\text{QM, Q}) - E(\text{MM, Q})] \quad (1)$$

that can be iterated to consistency.

Clearly, if we had sufficient computer resources, we would compute E(QM,all) directly by solving the time-dependent Schrodinger equation directly. Lacking that overall ability at the present time for large enzymatic systems, we can, however, if the quantum region size (Q) is modest, determine estimates for the three quantities on the right-hand side of Eq. (1). In the case of the ONIOM method, the boundary is chosen in practice by identifying single bonds in the side chains of residues that project into the proposed catalytic region but that are not expected to take part directly in the bond forming/breaking of the reaction. These bonds are then terminated with hydrogen atoms which are included in the calculation of E(QM,Q) and E(MM,Q). Since E(QM,Q) and E(MM,Q) include the electrostatic fields from atoms outside the Q region (defined as electronic embedding—in practice it becomes necessary to reduce the charges on atoms within several bonds of the Q region). The ONIOM method is flexible in that the critical catalytic region can be treated at a high quantum level, while the immediate surrounding region can also be treated at a lower level quantum region, thereby allowing for polarization. A recent application of the ONIOM (QM:MM) method by Ding, Chung, and Morokuma (2013) to a photochemically induced decarboxylation reaction of a green fluorescent protein illustrates the power and flexibility of the method. Developments leading up to this application are considered by Chung, Hirao, Li, and Morokuma (2012).

The major computer modeling packages—(AMBER (Wang, Wolf, Caldwell, Kollman, & Case, 2004), CHARMM (Brooks et al., 2009), GROMOS (Scott et al., 1999), and NAMD (Phillips et al., 2005))—all have QM/MM software with varying degrees of flexibility. Individual laboratories have also contributed novel approaches, for example, the pseudo-bond approach (Zhang, Lee, & Yang, 1999), the QTCP approach (Rod & Ryde, 2005), and a high-dimensional string QM/MM free energy method combined with an enhanced-sampling technique (Rosta, Yang, & Hummer, 2014).

A particularly insightful and recent review of the application of QM/MM methods to enzymes has been provided by Van der Kamp and Mulholland (2013) and a measure of how good certain variants of the method perform versus a full QM calculation can be found in a review by Hu, Soderhjelm, and Ryde (2011).

3. DNA POLYMERASE β

It is probably true that more crystallographic data have been collected on DNA polymerase (Pol) β than any other polymerase (Beard & Wilson, 2014). This 39-kDa DNA gap-filling enzyme has occupied much of the chemical, structural, and theoretical effort of our laboratories in recent years. The definitive structure for defining the active site for NTP insertion was published in 2006 (Batra et al., 2006) and was made possible by the use of a nonreactive NTP to impede the reaction. Previous approaches relied on using a dideoxy-terminated DNA substrate that lacked an atom that participated in catalysis, i.e., primer O3′ (Sawaya Prasad, Wilson, Kraut, & Pelletier, 1997). The comparative analysis of structures and kinetics of Asp256 (wild type) along with D256E and D256A variants (Batra et al., 2013) has established that Asp256 is the catalytic base for triggering the insertion reaction. Specifically, the transfer of a proton from the sugar O3′ to the catalytic base initiates the reaction. As described below, Pol β universally describes the action of nucleic acid polymerases.

3.1. Abashkin, Erickson, and Burt (2001): A cluster QM calculation at steps along a proposed reaction path

A brave and early attempt at an atomic level understanding the chemistry of the insertion step of Pol β was by Abashkin et al. (2001). This study, while not QM/MM, motivated our later work in that it focused closely on what was known from structural experiments at the time to define a realistic initial state for theory. In this case, the reference crystallographic structure (2.9 Å resolution) was determined by Sawaya et al. (1997). The theoretical study that followed was a set of DFT(double zeta)/QM calculations on an electrostatically neutral cluster of 67 atoms that defined various steps along two postulated reaction paths (Fig. 1).

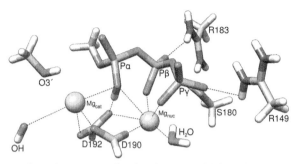

Figure 1 Proposed prechemistry molecular cluster model for Pol β (PDB = 1BPY; Sawaya et al., 1997). The model has many features revealed by higher resolution structures; however, these structures (Batra et al., 2006) exhibit an aspartate (residue 256) rather than OH- on the catalytic magnesium. In addition, the structures reveal a conserved water molecule coordinating this metal and further do not have a hydrogen ion on the metal-bridging oxygen of Pα. Abashkin et al. (2001).

A path with an intermediate PO3 was found to have an unreasonably high reaction energy barrier and so an alternate scheme involving a penta-coordinated transition state was explored. In this alternate scheme, the reaction initiates by the proton on the O3′ (modeled, since the X-ray structure was obtained without this group to avert reaction) jumping to the free oxygen on the α phosphate from which it is further transferred, after some adjustments, to an oxygen atom of the β phosphate of the departing pyrophosphate product. The original modeled position on the O3′ proton may be an essential element in determining much of what happens downstream. The QM energies of seven structures, including the initial and final minimized complexes, were determined. The product state was 19 kcal/mol more stable than the initial state. This is a reasonable result, but mostly based on the 1997 crystal structure. Several modifications to the crystallographic structure were made to arrive at a stable initial QM form which was electrostatically neutral: the necessary 3′-hydroxyl group was added, a hydroxide ion was added to the catalytic magnesium ion, a proton was added to Asp190, which helps bridge the two active site magnesium ions, and this proton was placed so as to hydrogen bond with the Pα–oxygen that also bridges the magnesium ions. Finally, a hydrogen bond was defined between the side chain of Arg149 and a γ-phosphate oxygen. The catalytic magnesium may be therefore somewhat undercoordinated and the hydrogen added to Asp190 has no structural basis (in fact, it was added for ensuring calculated stability of the initial structure).

3.2. Rittenhouse, Apostoluk, Miller, and Straatsma (2003): A cluster QM calculation on the prechemistry complex of Pol β

Although this study (Rittenhouse et al., 2003) did not map the reaction, it does highlight the details of the prechemistry active site. By building possible models of the active site based on the 1997 X-ray crystal structure of Pol β (Sawaya et al., 1997), an active site is proposed in which Pα-O, Asp190, and Asp192 bridge the two magnesium ions in a largely symmetrical fashion. In addition, a water molecule was proposed to be tightly bound to the catalytic magnesium ion. Asp256 is also proposed to form a tight bond with the catalytic magnesium ion. The three active sites aspartates are chosen to be -1 formal charge, while the NTP was set up with a -4 charge. The DNA and substrate (NTP) bases were included along with water molecules bound to each magnesium ion. However, the charge balancing arginines and the proton on the metal-bridging oxygen present in the Abashkin et al. (2001) study were not present. Thus, the overall charge of the cluster was -3. In their model, derived from quantum *ab initio* and DFT optimizations, the O3' proton has an orientation that is poised for transfer to the unbound negatively charged α-phosphate oxygen.

3.3. Radhakrishnan and Schlick (2006): QM/MM study of Pol β (based on PDB = 1BPY; Sawaya et al., 1997)

This study as well as work by Lin et al. (2006) that follows appear to be the first works to apply the QM/MM methodology to the Pol β system. The initial structures for the reaction are taken from earlier studies by the authors (Radhakrishnan & Schlick, 2004, 2005) that had focused on subdomain motions associated with NTP binding. Both the correct NTP (Pol β/DNA/dCTP for G:dCTP) and incorrect insertion (Pol β/DNA/dATP for G:dATP) reactions were considered. While based on the 1BPY structure (Sawaya et al., 1997), modifications were made to the active site so that three aspartates (190, 192, 256) are somewhat involved (precise geometries are not given, particularly for the units that bridge the two magnesium ions). The initial incorrect insertion structure was modeled from the initial correct insertion structure. The QM region including the link hydrogen atoms at the boundaries consists of 86 atoms: 7 atoms from each aspartate side chain (all assumed to bear -1 charge), 9 atoms from the 3' hydroxyl region of the primer sugar, 17 atoms from the NTP (assumed to bear -4 charge and terminate with a (link H)–CH_2–O–Pα unit), 7 atoms from the Ser180 side chain, 19 atoms from the Arg183 side chain (+1 charge), two magnesium ions (+2 charge), and 7 H$_2$O molecules (each Mg ion apparently has a water molecule bound in the initial structures). The charges add up to −2 although the charge was indicated to be −1. When using the 1BPY structure as the starting structural model, the 3′-hydroxyl group must be added; apparently, the location modeled for the proton on O3′ was such that the proton was not hydrogen-bonded to Asp256. A later high-resolution structure suggests this feature (PDB = 2FMS; Batra et al., 2006), where O3′ binds the catalytic magnesium (Mg$_{cat}$). The QM/MM calculations employed an existing interface between GAMESS-UK (Schmidt et al., 1993) for the QM calculations (6-311G basis) and CHARMM (Brooks et al., 2009) for the MM calculations. The reactions were studied by generating new structures by constrained MD on modeled structures with modified O3′–Pα and O3′–Mg$_{cat}$ distances. These new structures were subjected to energy minimization, and of these (50 total), four intermediates and a final product were found. These forms were additionally subjected to QM/MM dynamics to reach a total of six structures (reactant, four intermediates, and product) for both the correct and incorrect insertion reactions. For both reactions, the first intermediate is found to involve the O3′ proton jump to a water molecule and, although the paths of the proton after that are somewhat different for the two reactions, it ends up in the product structure on the γ-phosphate of the NTP. The transition barriers for a presumed bipyramidal transition state are estimated to be greater than 18 kcal/mol for the correct insertion and greater than 21 kcal/mol for the incorrect insertion, values that are consistent with those estimated from experimental studies (Ahn, Kraynov, Zhong, Werneburg, & Tsai, 1998; Berg, Beard, & Wilson, 2001). Due to the manner for generating intermediates, a reaction path construction was not possible. Shortly after this study, structural studies were published implying that the correct insertion initial system (PDB = 2FMS; Batra et al., 2006) has the −O3′–H proton strongly hydrogen-bonded to Asp256 and the incorrect insertion initial system (Batra, Beard, Shock, Pedersen, & Wilson, 2008) for G:dATP has O3′ displaced from Mg$_{cat}$ with a water molecule completing the catalytic metal hydration sphere. Both of these structural findings were accommodated in the next three studies discussed.

3.4. Lin et al. (2006): QM/MM study of Pol β (based on PDB = 2FMS; Batra et al., 2006)

New structures of Pol β indicated that the catalytic metal site can be occupied by sodium or magnesium ions (Batra et al., 2006). Refinement of the latter magnesium structure was at 2.0 Å. The major change from the 1997 lower resolution structure was that the catalytic metal was now octahedrally coordinated, with clear density for all water ligands, Asp256 and the O3′ oxygen of the primer terminus sugar (Fig. 2).

After performing cluster calculations (DFT/B3LYP) on several models suggested by the 2FMS structure, and motivated by the earlier DFT/QM study of Abashkin et al. (2001) and by the QM study of Rittenhouse et al. (2003), Lin et al. (2006) were able to map a stable QM/MM reaction path (not shown) that did not involve a proton on the bridging Pα oxygen. Arginine residues 183 and 149 also were not included. Asp256, which is bound to the catalytic metal, serves as the catalytic base for transfer of the O3′ proton. Separate QM/MM calculations to determine the most stable position of the O3′ proton for the cluster found that it was located between O3′ and an Asp256 oxygen atom as the central part of a hydrogen bond. The initial geometry of the O3′ proton is consistent with the O3′–OD2–Asp256 distance (2.81 Å) of the 2FMS structure. Both metals remain six coordinate throughout the reaction. To gain the effect of the entire protein, QM/MM

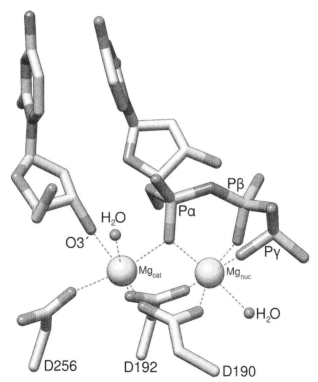

Figure 2 High-resolution structure (PDB = 2FMS; Batra et al., 2006) of the prechemistry complex of human Pol β. Both magnesium ions are octahedrally coordinated.

calculations were performed using the ONIOM method (with electrostatic embedding) discussed in Section 2 (Lin et al., 2006). The QM region consisted of three water molecules, the NTP with a proton on the γ-phosphate and a link atom after the O3′ sugar oxygen, three aspartate residues with link atoms between the β- and α-carbons, and the primer sugar with a link atom at the exit from the sugar ring (Fig. 3) for a total of 64 quantum atoms. In the crystallographic structure (Batra et al., 2006), there is an apparent hydrogen bond linking the O3′ proton to an oxygen of Asp256, and this oxygen, along with O3′, is also coordinated to the catalytic metal. This hydrogen bond telegraphs the transfer of the O3′ proton to Asp256 which activates the nucleophile (O3′ anion) for attack on the α-phosphate. The cluster study of Abashkin et al. (2001), based on a less detailed X-ray crystal structure, especially near the catalytic metal, had O3′ proton initially transferred to the α-phosphate.

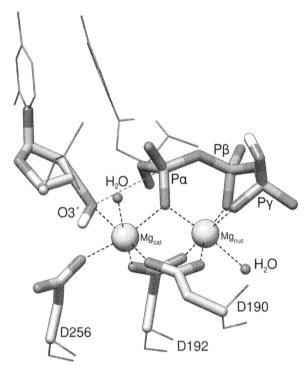

Figure 3 Model of the Pol β prechemistry active site based on PDB = 2FMS (Batra et al., 2006). A water molecule is shown on each magnesium ion. A third water molecule is not shown for clarity. *Lin et al. (2006).*

The initial equilibrium structure (before insertion) was obtained by adding protons to the 2FMS structure (Batra et al., 2006). The nonhydrolyzable analog, 2′-deoxy-uridine-5′(α,β)-imido triphosphate, located in the active site of the X-ray crystal structure was changed to dTTP to facilitate the reaction. Protonation states of amino acids were set at pH 7.0 via http://propka.chem.uiowa.edu. All crystal waters were preserved. Water and counter ions were added to provide a box that included 21,367 water molecules and 25 sodium ions and were electrically neutral. The SANDER module of the Amber 8 package (Case et al., 2005) with the Amber ff99 force field (Wang et al., 2004) was employed for dynamics simulations and minimizations and the particle-mesh Ewald code (Darden et al., 1993; Essmann et al., 1995) was used for long-range electrostatic interactions. The ONIOM module implemented in Gaussian 03 (Vreven et al., 2006) was the base QM/MM method with electronic embedding adopted for the study.

The strategy was to map the energy as a function of two variables: the forming O3′–Pα bond and the breaking Pα–O(Pβ) bond. Thus, using the scan keyword in Gaussian 03, a map of the energy versus these two variables was generated using the quantum method/basis set of B3LYP/6-31G*. In the early stages of the reaction, the proton on O3′ transfers to the Asp256 carboxylate group with a low barrier of about 3.5 kcal/mol. Once this step occurs, the O3′–Pα distances close until a transition state (with no stable intermediate) occurs. The geometry at the transition state is defined by R-O3′–Pα = 2.2 Å and R-Pα–O(Pβ) = 1.9 Å. The transition state barrier above the initial reactant state is 21.5 and 18.0 kcal/mol above the deprotonated intermediate state. These barriers are consistent with experimental estimates of a free energy of activation of 16 kcal/mol (Berg et al., 2001). At the last point along the reaction path, the product state is −5.2 kcal/mol below the initial reactant state and the reactant bond has now closed to R-O3′–Pα = 1.65 Å and the broken bond has expanded to R-Pα–O$\alpha\beta$ to 3.27 Å.

An electrostatic energy decomposition study was undertaken at the transition state to ascertain which amino acids were stabilizing and which were destabilizing. The energy of interaction was taken to be

$$E_{\text{QM/MM}}^{\text{esp}} = \sum_{\substack{\alpha \in QM \\ \beta \in \text{residue}}} \frac{Q_\alpha Q_\beta}{R_{\alpha\beta}}$$

Then, the difference between the residue's position at the transition state and at the initial deprotonated state (i.e., O3′ deprotonated early in the path) is

$$\Delta E = E^{\text{esp}}(\text{residue, TS}) - E^{\text{esp}}(\text{residue, initial state})$$

The two residues that contributed the most to stabilizing the transition state were Arg149 and Arg183 (-4.6 and -7.1 kcal/mol, respectively). These stabilization energies are in accord with their vicinity to the pyrophosphate leaving group in the initial state (modeled from the X-ray crystal structure) and so, being charged themselves, they can modify the response of the breaking bond as the reaction proceeds.

A central observation of this QM/MM study was the remarkable stability of the geometry of the two magnesium ions, both of which interact relatively symmetrically with an O–Pα atom, which together form a scaffold upon which the reaction appears to evolve.

3.5. Lin et al. (2008): Pol β incorrect insertion QM/MM study (based on PDB = 2C2K; Batra et al., 2008)

The appearance of a Pol β mismatch structure (G:dATP) (Batra et al., 2008) provided a new opportunity for understanding of misincorporation at the atomic level. Experimentally, the misincorporation step is experimentally estimated to be >600-fold slower for G:dATP than G:dCTP (Ahn et al., 1998; Bakhtina et al., 2005), but it does occur and is measurable. Examination of the X-ray crystal structure PDB = 2C2K of this mismatch (Batra et al., 2007) suggests why. The O3′ of the primer terminus is no longer coordinating the catalytic metal; it has been displaced by a water molecule. A direct path was tested to determine if O3′ would bond with Pα after the O3′ proton transferred to the Pα-free oxygen. The energy barrier was found to be very high for this path (i.e., 48 kcal/mol), suggesting that another path must instead be viable. Another alternative was to instead propose a two-step mechanism, in which the intrusive water molecule was synchronously moved from the magnesium coordination at the same time O3′ gained coordination. Constrained molecular dynamics was used to force this transfer and provide a prechemistry state that had O3′ strongly interacting with the catalytic metal. The resulting structure, found by equilibration molecular dynamics, was very similar to the Lin et al. (2006) prechemistry structure. The barrier for this process was found to be about 14 kcal/mol from a B3LYP/6-311G** calculation on a cluster that included key coordinations for the two metals. The O3′–Mg$_{cat}$ distance served as the driving coordinate for the transformation. The idea was that the ground-state structure, derived from careful equilibration of the X-ray crystal structure by molecular dynamics, and which had a high reaction barrier, underwent at conformational change (O3′ moves to displace the Mg$_{cat}$ coordinating water) that costs 14 kcal/mol, but which created a prechemistry structure from which the reaction path energetics could be determined from QM/MM procedures as in Lin et al. (2006). The energy barrier, using QM/MM procedures with the ONIOM method similar to Lin et al. (2006) for the misinsertion step, is about the same as found for the correct insertion. The ratio of the insertion rates (correct/incorrect) is 12.5/0.019, which can be converted to an energy difference of 3.8 kcal/mol, and then interpreted as the difference between the ground states of G:dCTP versus G:dATP. In this view, for correct insertion, the ground state is the prechemistry state, while for incorrect insertion, the ground state is separated thermodynamically from the prechemistry state by 3.8 kcal/mol and by a nonrate-limiting barrier of 14 kcal/mol (water ↔ O3′ switch). The switch must occur before reaction can occur. Electrostatic effects of residues lying outside the quantum region on the transition state energies were found to be similar to those found for the correct insertion study.

3.6. Batra et al. (2013): A QM/MM study based on the X-ray crystal structure of the Pol β variant D256E

The two previous studies (Lin et al., 2008, 2006) concluded that the catalytic base for the deprotonation of the O3′ proton of the primer terminus was the oxygen atom of Asp256 that is also bonded to the catalytic magnesium ion. In order to test this idea further, two variant structures (Fig. 4) were determined: D256E and D256A (PDB=4JWM and 4JWN, respectively). The D256E structure (PDB=4KWM) is charge conservative but significantly different than the D256 structure (PDB=2FMS) in two important ways: the carboxylate side chain is not bonded to the catalytic metal as in wild type (the coordination position on the catalytic metal is now occupied by a water molecule) and Arg254, which helps anchor the side chain of Asp256 in place in the wild type, no longer interacts with the 256 position carboxylate. The 3′ oxygen is, however, still in essentially the same place as for the wild-type (PDB=2FMS) structure, approximately 3.5 Å from the Pα of the NTP. Also, for both the D256E and wild-type structure, there exists an apparent hydrogen bond between the O3′ proton and a carboxylate (256 position) oxygen. The change is more drastic for the D256A

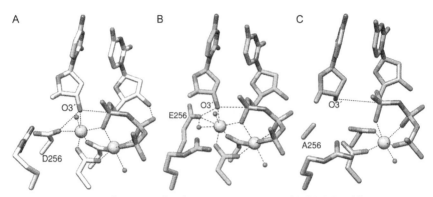

Figure 4 X-ray crystal structure for the active site region of Pol β. (A) Wild-type enzyme: PDB = 2FMS (see Fig. 3). (B) D256E variant (PDB = 4JWM). A water molecule occupies the vacated Asp256 oxygen position. The alignment of O3′–Pα–Oαβ is, however, similar to the wild-type system. The insertion reaction rate is reduced nearly three orders of magnitude. (C) D256A variant (PDB = 4JWN). The catalytic metal of (A) and (B) is lost in D256A variant. The insertion reaction is very slow but can be partially recovered at high pH.

system—the structure (PDB = 4JWN) does not have the bimetallic active site, but instead has only one magnesium ion which is located in the nucleotide binding position.

The catalytic metal site is empty (Fig. 4) and the 3′ oxygen is displaced to 4.9 Å from the Pα of the NTP as compared to about 3.5 Å in the 2FMS wild-type structure. Kinetic measurements of k_{pol} of the insertion rate indicate that the rate of insertion for the D256E system was reduced by three orders of magnitude as compared to the wild type, while the D256A system had no measurable activity at pH 7.4. Increasing the reaction pH recovered significant activity (~10-fold/pH unit), indicating that the required deprotonation event had a pK above 10. QM/MM reaction paths, using a QM region, were similar to those of the Lin et al. (2006) study. In addition, the basis set was now 6-31G*, an extra –CH$_2$ group was included in the QM region on the aspartates, and only the O3′–Pα distance was mapped using the electronic embedding methodology of the Gaussian 09/ONIOM interface (Case et al., 2010; Vreven et al., 2006) for both the wild type (a QM/MM study was included for internal reference) and the D256E variant systems. Because the alignment of O3′–Pα–Oαβ is similar to that of the wild-type system, preprocessing to obtain a prechemistry state was not necessary. To test the location along the reaction path of the O3′ deprotonation, two paths were investigated: one where O3′ deprotonates early and one where it deprotonates late, i.e., near the transition state. The former case occurs for the wild type (as in the Lin et al., 2006 study) and the latter case is found for the variant form. The transition state barriers are 14 and 21 kcal/mol for the wild type and variant cases, respectively, while the O3′–Pα distance at the transition state is the same for both cases. The higher energy barrier for the variant is consistent with the greatly lowered experimental insertion rate observed for the variant. A charge analysis using the Merz–Kollmann (Besler, Merz, & Kollman, 1990) option of Gaussian 09 showed that a more effective nucleophile (O3′ anion) is developed at the transition state for the wild type as compared to the variant system. An electrostatic energy decomposition analysis comparison between the wild type and D256E variant did not reveal a clear reason as to why the variant transition barrier is higher; however, when the electronic embedding of the ONION procedure was turned off, thereby minimizing the transition state lowering effects of residues outside the quantum region, the barrier increase was much greater for the variant (increased from 21 to 58 kcal/mol) than for the wild-type enzyme (increased from 14 to 42 kcal/mol). Thus, the location of the Glu256 side chain in the variant D256E is more destabilizing overall relative the location of the Asp256 side chain in the wild-type system, even though both systems are aligned well for the reaction and both have an apparent hydrogen bond for the O3′ proton with the carboxylate of the catalytic base. This particular paper is especially interesting because it combines structure determination, kinetic studies, and theoretical estimation of a reaction path to investigate the nature of the chemistry step of the Pol β system.

3.7. The broader picture: Pol β

Some DNA polymerases such as Pol β undergo subdomain conformational changes when the NTP substrate binds, whereas others (e.g., Pol λ) do not (Wu et al., 2014). When the conformational change is present, the question arises as to whether these prechemistry events are kinetically and/or thermodynamically influenced by the overall mechanism. In addition, there are related questions as to what property (or properties) of the system control substrate discrimination (i.e., fidelity). Theoretical viewpoints differ somewhat, as seen by the attempt by Mullholland, Roitberg, and Tunon (2012) to adjudicate a lively discussion between Schlick, Arora, and Beard (2012) and Prasad, Kamerin, Florian, and Warshel (2012) about the relative contributions to the mechanism of prechemistry conformational changes to the observed barrier for catalysis. Further insight into the mechanism of DNA polymerases, if not resolution of conflicts, was provided by a commentary by Tsai (2014) and the experimental work (Olson, Patro, Urban, & Kuchta, 2013) that showed that for the free energy difference between correct and incorrect insertion (averaged over two families and many substrates) was of the order of 5 kcal/mol. This result would imply that the moderate fidelity of the Pol β may simply be due to this large thermodynamic difference, although it does not tell us about the underlying molecular details. Finally, the relationship of Pol β's mechanism of action to accumulating structural data, kinetics, and computation has been summarized recently (Beard & Wilson, 2014).

4. APPLICATION OF QM/MM TO SYSTEMS SIMILAR TO POL β

Five of the major DNA polymerase families (A, B, C, X, and Y) were recently compared structurally for similarities (Wu et al., 2014) with a focus especially on the geometry of the active site of the ternary complexes (polymerase, double-stranded DNA, and nucleoside triphosphate) that lead to the chemistry of the formation of the O3′–Pα bond and the breaking of the Pα–Oαβ–Pβ bond to form pyrophosphate. Essential elements that appear to span these five families in the active site are the existence of two negatively charged aspartate side chains that bridge two divalent metal ions that are separated by 3.5–4.0 Å with nothing directly between these highly charged ions, the full NTP unit, the primer terminal O3′. Another essential element appears that for optimal function, including Watson–Crick insertion fidelity, the two divalent metals ions should be magnesium ions. Given this background, let us consider three recent QM/MM studies, one of which considers chemical insertion in another member of the X-family (Pol λ) and two of which consider chemical insertion in members of the Y-family.

4.1. Pol λ insertion

For Pol λ (X-family DNA polymerase), insertion (Cisneros et al., 2008) is a QM/MM study based on a published X-ray crystal ternary substrate complex structure (PDB=2PFO; Garcia-Diaz, Bebenek, Krahn, Pedersen, & Kunkel, 2007) which had many features similar to the structure of Pol β (PDB=2FMS). The X-ray crystal structure (Fig. 5) consisted of the precatalytic complex of double-strand DNA, Pol λ, and a nonhydrolyzable NTP (dUMPNPP) served as the starting template for a QM/MM study of the chemical insertion reaction. To prepare the system, the Mn(II) ion in the catalytic metal site was replaced by a magnesium ion and the dUMPNPP

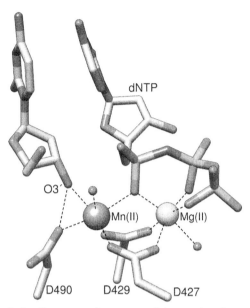

Figure 5 High-resolution X-ray crystal structure of the prechemistry complex of human Pol λ (PDB=2PFO). The catalytic site is occupied by a manganese ion, whereas the nucleotide binding site hosts a magnesium ion. Two water molecules coordinating the metal ions are also shown. *Garcia-Diaz et al. (2007)*.

was replaced by dUTP, hydrogen atoms were added, and the system was solvated in a large box of water and equilibrated with a 2-ns PMEMD simulation (Case et al., 2005). All atoms within 30 Å of the catalytic metal were retained for the starting system. The QM/MM calculations were performed with the pseudo-bond method (Zhang, 2005; Zhang et al., 1999) which employed a modified version of Gaussian 03 (Frisch et al., 2004) with TINKER (Ponder, 1998) to compute energies along the reactions paths studies. Reaction coordinates for the proposed paths were chosen as described (see below) in Cisneros et al. (2008). A QM subsystem was chosen that consisted of the NTP through the C5′ sugar atom, side chains of Asp490 (the equivalent of Asp256 in Pol β) and Asp427 and Asp429 (Asp190 and 192 in Pol β), the two magnesium ions, part of the primer sugar terminus (excluding the phosphate and C5′), and two metal-bound water molecules for a total of 72 atoms (Fig. 5). The boundary atoms defining the pseudo-bond locations are the aspartate Cαs, C5′ of the primer dC and C4′ of the incoming NTP. After protonating the γ-phosphate, the NTP had a formal charge of −3 and the QM region a net charge of −2. The QM method was B3LYP (Becke, 1993; Lee, Yang, & Parr, 1988) with a combined basis of 6-31G* for atoms involved in the paths proposed and the 3-21G for the nonreactive atoms, and the LANL2DZ pseudo-potential (Wadt & Hay, 1985) was employed for Mn (paths in which the catalytic metal was either Mn^{2+} or Mg^{2+} were investigated). An extra diffuse function was included on Pα to accommodate Pα hybridization changes. The techniques used to produce the reaction path coordinates (Cisneros et al., 2008) required a product state structure; this was produced from the reactant state using modeling and QM/MM optimization. This nonexperimental structure, which anchors the product end of the reaction path, thereby interjects a degree of uncertainty into the process. Once a given test reaction path is specified, a reaction coordinate can be defined. Unfortunately, the equations defining the reaction coordinates for the reaction paths are not stated explicitly. In this study, the several paths (with magnesium in both sites) that were investigated were initiated by (i) transfer of the O3' proton to Asp490, the analog of Asp256 in the Pol β system; (ii) the transfer of the O3' proton to Asp429, one of the metal-bridging aspartates; (iii) transfer of the O3' proton to one of the water molecules bound to the metal ions; and (iv) transfer of the O3' proton to the free oxygen on Pα. All of the paths, except the first, which transfers the O3' proton to the nonbridging Asp on the catalytic metal, have high-energy transition states. The systematic generation of the coordinates along the reaction path (equilibrated experimental reactant structure) to (equilibrated generated *in silico* product structure) permitted the determination of the energy versus reaction coordinate profile for the systems with either magnesium or manganese ions at the catalytic metal site. These profiles for both metal ions gave approximately the same value for the activation energy ($\sim$17 kcal/mol) with the suggestion of a weakly bound intermediate between two transition states. It may be that the intermediate seen is an artifact of the constraint imposed by the method of defining the reaction coordinates. The distances between the metals change very little over the path (3.5 Å for Mg–Mg and 3.7 Å for Mn–Mg). Overall, despite the fact that different QM/MM methods were employed, the conclusions about the path of the reaction and magnitude of the activation energy determined are similar to that found in the QM/MM study for the wild-type Pol β system (Batra et al., 2013).

4.2. Dpo4 insertion

The Dpo4 (Y-family DNA polymerase) insertion reaction (Wang & Schlick, 2008) was a QM/MM study based on the X-ray crystal structure of the Dpo4/DNA complex of 8-oxoG:dCTP (PDB=2ASD; Rechkoblit et al., 2006). The X-ray crystal structure has calcium ions in the active site and O3' is missing by design to stop insertion during structure determination. Thus, this structure is somewhat distorted; for instance, the modeled O3'–Mg$_{cat}$ distance is 5.3 Å (~3 Å too long), the modeled O3'–Pα distance is 4.7 Å (~1 Å too long) and the Me–Me distance is approximately 0.8 Å too long. The distorted X-ray crystal structure was then modeled to be similar to the high-resolution Pol β structure (Batra et al., 2006; PDB=2FMS), although it is not clear if the coordination state of the NTP-coordinating Mg ion is complete or the distance between the active site metals. The quantum part of the QM/MM was chosen to include the two bridging aspartates, a Mg$_{cat}$ coordinating glutamate, a small part of the primer terminus, two modeled magnesium ions, the incoming dCTP, and four water molecules that coordinate Mg$_{cat}$ and dCTP. The total charge on the quantum system is apparently -3 as the dCTP is fully charged (-4). Once the initial prechemistry state was established, several possible paths were investigated. The QM/MM procedure was similar to the earlier study on Pol β by Radhakrishnan and Schlick (2006). The boundary between QM and MM is handled by a link hydrogen atom. The assumption was that the O3' must initially be deprotonated before the O3'–Pα bond can form and the Pα–Oαβ–Pβ bond can break. Given this, the paths tested were to transfer the proton from O3' to (i) one of the water molecules, (ii) the free oxygen on the Pα, (iii) an oxygen on Glu108, which is in the same position as Asp256 in the Pol β structure, or (iv) one of the bridging aspartates. The best energy path was concluded to be the transfer of the proton to water molecules with a subsequent transfer to the γ-phosphate. Three intermediates are found along the path to products (Pα–Oαβ–Pβ bond broken, O3'–Pα bond formed) with the highest energy intermediate one in which the proton is transferred through water molecules to the free oxygen on Pα. Although a transition state for this transfer is shown (20 kcal above the initial state and 5 kcal above the intermediate), a description of how this curve was determined and subsequent transition states are not given. The overall conclusion is that the rate-limiting step (O3' proton transferred to water to Pα oxygen) occurs about 20 kcal above the reactants.

4.3. Pol κ insertion

Pol κ is also a Y-family polymerase that bypasses certain lesions such as benzopyrene. This study by Lior-Hoffmann et al. (2012) employs the pseudo-atom method (Zhang et al., 1999). A ternary substrate complex X-ray crystal structure (PDB = 2OH2; Lone et al., 2007) is employed. This structure, which is missing the O3′ and does not have a catalytic ion, requires significant modeling to obtain a structure suitable for defining an adequate prechemistry system. The modeled system involved establishing octahedral coordination at each magnesium ion. A backbone oxygen of Met108 occupies the sixth position of the NTP Mg_{cat} and Glu199 occupies the same position as is occupied by Asp256 in Pol β. A reaction coordinate driving procedure which proceeds by "stepping along a proposed reaction coordinate and performing energy minimizations with respect to the remaining coordinates" is employed (Hu & Yang, 2008; Zhang, Liu, & Yang, 2000). It is not clear what the actual reaction coordinates used are in this work, but generally they appear to partially describe the transfer of protons. Also, not all of the geometric parameters are defined. The choice as to which atoms to include in the QM part of the QM/MM scheme is somewhat different than for the Dpo4 system. For the Pol κ system, 81 atoms are included in the QM part: the NTP, two water molecules (one of which resides on Mg_{cat}), the two Mg ions, the primer terminal sugar and base, and the side chain of Glu199. The two metal ion bridging aspartates and the backbone atoms of Met108 (which coordinates the Mg_{nuc}) are relegated to the MM subsystem. The NTP is taken to be fully charged (−4) so that the total system charge is −1. Several reaction paths for the deprotonation of the modeled O3′ proton were tested: (i) employing Glu199 as a catalytic base or (ii) employing the free oxygen on Pα as the catalytic base. These potential paths were rejected when the protonated species appeared to be unstable for short interspersed QM/MM–MD simulations. The path (reaction paths tested are not precisely defined) that was found to be desirable was (iii) a transfer of the O3′H proton to the γ-phosphate through the two water molecules included in the QM subsystem leading to a stable intermediate, followed by the transfer of this proton to the β-phosphate as the O3′–Pα bond formed and the Pα–Oαβ–Pβ bond broke. Similar to the Dpo4 system, the initial proton transfer step was found to be rate limiting. The free energy of activation was found to be approximately 11 kcal/mol.

These last two QM/MM simulations discussed (Dpo4 and Pol κ), although using different ways of handling the boundary between QM/MM, share similarities. Both of the reference X-ray crystal structures are significantly distorted, are missing the O3′ unit, and do not have a Mg–Mg ion pair at the core of the active site until modeled. Additionally, neither paper gives geometry details about the modeled active site. For instance, it is impossible to deduce from these papers the distances from O3′ to the oxygen atoms of the glutamate that occupies the Asp256 position in Pol β in their prechemistry structures. If the Lin et al. (2008, 2006) and Batra et al. (2013) QM/MM papers are correct for the mechanism of Pol β and other similar systems, the reaction of the chemistry insertion is initiated by the transfer of the O3′ proton to the side chain of Asp256 that is nearest O3′ and together Asp256 and O3′ are bound to the catalytic magnesium. Both the Dpo4 and Pol κ papers consider the possibility of this mechanism: for Dpo4, the transfer of the O3′ proton to Glu108 and for Pol κ, the transfer of the O3′ proton to Glu199. In both cases, the cause for rejecting this path is stated to be that the transfer state is found unstable. In both cases, however, not enough detail (which carboxylate oxygen for transfer was tested, what was the distance from the O3′ to this oxygen in the modeled prechemistry structure, what was the precise geometric system tested for stability) is given to be assured that this path (O3′ to glutamate) was thoroughly vetted. This issue becomes important later.

5. RNA POLYMERASE

The understanding of the mechanism of the synthesis of mRNA was greatly advanced by the publication of a group of structures of RNA Polymerase II by Wang, Bushnell, Westover, Kaplan, and Kornberg (2006) derived from *Saccharomyces cerevisiae*. One of these structures, PDB=2E2H, was chosen by Carvalho, Fernandes, and Ramos (2011) as the basis of a QM/QM study on the mechanism of extension of mRNA by Pol II. The structure (resolution 3.95 Å) does not have an O3′ on the ribose primer and Pα of the GTP substrate is not perfectly seated between the two Mg ions (as in the case of higher resolution structures of DNA polymerases). The Mg ions are 3.43 Å apart which is consistent with known DNA polymerase structures. While Asp485 is firmly attached to the catalytic Mg ion and two aspartates serve as bridges between the two magnesium ions, the resolution is insufficient to observe a complete coordination shell about the magnesium ions. The ONIOM/Gaussian 03 (Vreven et al., 2006) method is used for optimizing trial structures generated to satisfy trial paths chosen for investigation. A QM/QM procedure in which there is a higher level QM (DFT B3LYP/6-31G(d)) applied to inner core atoms and a lower level QM (PM3MM, Stewart, 1989) applied to an outer shell of atoms. The ONION/Gaussian 03 programs allow both QM/MM and QM/QM procedures. Final energies of the optimized structures were then computed at the B3LPY/6-311++G(2p,2d) level. A neutrally charged model of the active site was extracted (Fig. 6) that consisted of parts of four aspartates, GTP, the primer terminal ribose, the two Mg ions, parts of a histidine, a lysine, and three arginines for a total of 226 atoms. For the inner core, the two Mg ions, GTP, the ribose, histidine, and last three side-chain atoms of the four aspartates were chosen. There apparently were no water molecules included in the region of quantum calculation. After preparing the models by adding the missing O3′ to the ribose and missing H atoms, and relaxing in neutralized water for 20 ns, four hypothetical reaction pathways were generated. These were (i) HYP1 (the O3′ proton jumps to an α phosphate oxygen on the GTP and ends up protonating the pyrophosphate after bond forming (O3′–Pα) and bond breaking (Pα–Oαβ)), (ii) HYP2 (a hydroxide ion materializes near the 3′-hydroxyl group, deprotonates O3′ so that it can attack Pα, and the departing pyrophosphate product is stabilized by a proton transfer from a nearby histidine), (iii) HYP3 (an OH⁻ group is added on the catalytic magnesium ion initially, which deprotonates the O3, while the product pyrophosphate is stabilized by a proton from the histidine); and (iv) HYP4 (the nonbridging aspartate bound to Mg$_{cat}$ accepts the proton from the 3′-hydroxyl group so that the O3′ anion can attack Pα and the histidine protonates the leaving pyrophosphate). The energy cost of

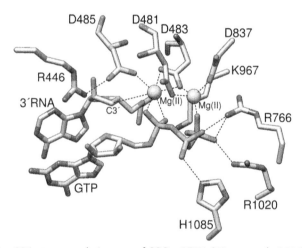

Figure 6 The 226 atom catalytic core of PDB=2E2H (Wang et al., 2006) chosen for modeling the RNA Pol II insertion reaction with the QM/QM procedure of G03/ONIOM (Carvalho et al., 2011). The higher lever quantum region is further reduced (not shown) to the triphosphate of the GTP, the RNA ribose, the side-chain terminal three atoms of each of the four aspartates, the two magnesium ions, and the His1085 side chain.

generating a hydroxide from bulk water for the HYP2 path was rationalized to be 7.5 kcal/mol by the use of free energy perturbation theory and concentration in bulk considerations. Only HYP2 (path 2) had a low activation energy (~10 kcal/mol), the other three paths had barriers of >29 kcal/mol. In this path, a hydroxide is created near O3′, which facilitates deprotonation of O3′. Then, the positively charged His1085 loses its proton to one of the free oxygens on Pβ. The consequence of this is the weakening of the Pα–Oαβ bond as the Pα–O3′ bond forms. It is this last step that defines the limiting reaction step in the overall path. If it is the case, as we speculated earlier, higher resolution structures ultimately lead to an initial active site that has what we think are the essential features (O3′ in place, an Asp/Glu group at the Asp256 Pol β position, and a water molecule coordinating Mg_{cat}, the Pα–O symmetrically (approximate isosceles triangle) bridging the two magnesium ions and there are two metal-bridging aspartates); this mechanism will require reinvestigation. Indeed, in the same paper that PDB=2E2H originates, one can also find PDB=2E2J. In the latter structure, O3′ is present (a nonhydrolyzable NTP is used) and it is tightly bound to the Mg_{cat}. In this structure, the aspartate coordinating to the Mg_{cat} is within H-bonding distance (2.52 Å) of the naturally present O3′. A later structure (PDB=3S1Q) has appeared (Liu, Bushnell, Silva, Huang, & Kornberg, 2011) from the same group, where the aspartate at Mg_{cat} exists and superposition of Pα–O and the two magnesium ions with PDB=2FMS is almost perfect even though O3′ is missing in the former structure.

Very recently, the insertion pathway for RNA Pol II has been further investigated (Zhang, 2013) in the Salahub lab at the University of Calgary. Several starting systems were considered: (i) model A, based on PDB=2E2H (Wang et al., 2006); (ii) model B, based of PDB=2E2J (Wang et al., 2006; Fig. 7); (iii) models C1 and C2 where, for both of these models, the NTP(GMPCPP) of PDB=2E2J was changed to the GTP of PDB=2E2H and optimized and the system equilibrated for either 1 ns (Model C1) or 12 ns (Model C2). The sizes of the various structures were reduced by fixing all atom coordinates beyond 20 Å of the NTP Pα. The quantum subsystem included parts of side chains of four aspartates (481, 483, 485, 837), three arginines (446, 766, 1020), three water molecules, the entire GTP or NTP substrate, the ribose of the primer terminus, and the two magnesium ions for a total of 144 atoms and charge of −1 for the quantum region. An in-house QM/MM program was employed that utilized hydrogen link atoms at the boundary. The MM part of the calculation was performed with the CHARMM27 force field (Foloppe & Mackerell,

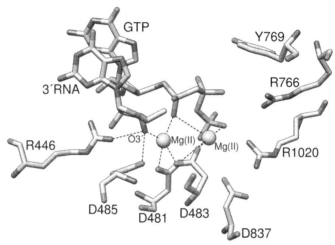

Figure 7 The catalytic core of PDB=2E2I chosen by Zhang (2013) for modeling the insertion of a GTP by RNA Pol II.

2000; MacKerrell et al., 1998), and the QM part of the calculation was performed with the semiempirical AM1/d-PhoT method (Nam, Cui, Gao, & York, 2007). The four models were subjected to the defined QM/MM procedure with the apparent conclusion that model C is most appropriate and that direct transfer of the O3′ proton to the α-phosphate will be the lowest energy path to products. We note that a path with initial transfer of the O3′ proton to Asp485 (the structural equivalent of Asp256 for Pol β), the side-chain carboxylate oxygen of which is 2.52 Å from O3′ in PDB=2E2J, was not investigated.

6. THOUGHTS FOR FUTURE QM/MM SIMULATIONS ON TERNARY SUBSTRATE COMPLEXES OF NUCLEIC ACID POLYMERASES

Has the experimental structural evidence that has accumulated to date reached the critical amount needed to be able to develop a more unified, consensus view of how the two metal site at the core of polymerase functions? The diverse mechanisms presented in the various QM/MM applications to polymerases that are presented in this manuscript reflect the lack of a unified view. However, we now believe that a solid case can be made for the "O3′ → initial proton transfer to an active site acidic residue" as the initial step in polymerase activity. For Pol β, this is Asp256. The data for this suggestion are collected in Table 1.

These data appear consistent with the proposal that if the crystal structure is performed with an nonhydrolyzable NTP substrate, so that O3′ is present, and if the active site contains two magnesium ions, then O3′ appears to be within a distance characteristic of a strong hydrogen bond in its interaction with the Lewis base occupying the Asp256 (Pol β) position. We note that the data in Table 1 span several nucleic acid families. These observations support the idea that the essential active site for DNA/RNA polymerase activity may involve two closely spaced magnesium ions, an NTP, two metal-bridging aspartates, and an O3′ that coordinates with the catalytic metal and with a Lewis base bound to the catalytic metal. In the higher resolution structures in Table 1, there is also a water molecule bound to the catalytic magnesium.

Table 1 Proposed O3′ proton transfer distance to a hydrogen-bonded base oxygen in X-ray crystal structures that have active site magnesium ions, a primer O3′, and a water molecule bound to Mg_{cat}

System	PDB entry	Resolution (Å)	O3′H–O(Asp/Glu) (Å) in PDB	AA at "D256 position"
Pol β	2FMS(c)	2.00	2.81	D256
Pol λ	2PFO(c)	2.00	2.61	D490
Pol μ	4M04(c)	1.90	2.81	D418
Pol η	3MR2(c)	1.83	2.67	E116
Pol β(a)	4JWM(c)	2.00	2.60	E256
RB69(b)	3SPY(c)	1.88	2.65	Water-6
RNA Pol II(c)	2E2J(c)	3.50	2.52	D485 RNA/DNA

A nonhydrolyzable NTP substrate is employed in all entries to permit inclusion of O3′.
(a) D256E variant: Glu 256 is not bound to the catalytic metal in the X-ray structure but is displaced by a water molecule. A potential hydrogen bond exists between O3′ and Glu256. The proton jump from O3′ to E256 takes place just before the transition state, whereas this jump for the wild-type enzyme (O3′ to Asp256) takes place far from the transition state.
(b) Four nonactive site mutations (L415A, L561A, S565G, and Y567A) were made to wild-type polymerase for structure determination. RB69 is a B family of DNA polymerase. A structure with O3′ and two magnesium ions in place has not been solved with the wild-type protein.
(c) 2FMS: Batra et al. (2007); 2PFO: Garcia-Diaz et al. (2007), Mn^{2+} in the catalytic site and Mg^{2+} in the nucleotide binding pocket; 4M04: Moon et al. (2014); 2MR2: Biertumpfel et al. (2010); 4JWM: Batra et al. (2013); 3SPY: Xia, Wang, Blaha, Konigsberg, and Wang (2011); 2E2J: Wang et al. (2006). The O3′–Mg_{cat} distance is 2.04 Å, Mg_{cat}–OD1485 = 3.7 Å, the nonhydrolyzable NTP not well seated ($-O \rightarrow CH_2$). The resolution is too poor to determine if Mg_{cat} coordinates a water molecule. It is possible that given the tight O3′–256O bond (2.52 Å), the long 256O–Mg_{cat} distance (2.93 Å), and the short O3′–Mg_{cat} distance (2.04 Å), that the proton on O3′ has transferred to a nearby oxygen (aspartate/glutamate) in the crystalline state.

If any of these features is missing and must be modeled, the modeling must be carefully done to restore all essential geometry for function. *One reasonable way to interpret the data in Table 1 is with the assertion that the insertion chemistry is triggered by the proton transfer of the O3′ proton through its hydrogen bond to the catalytic base (an aspartate or glutamate). We propose that this jump may take place by QM tunneling.* Proton tunneling is now a well-established concept for enzyme reactions (Hay & Scrutton, 2012; Klinman & Kohen, 2013; Klippenstein, Pande, & Truhlar, 2014; Kuznetsov & Ulstrup, 1999; Truhlar, 2010; Truhlar et al., 2004). Once this jump occurs, "the cat is out of the proverbial bag." Instantaneously, there is a reorganization of electronic charge involving at a minimum the proton, the catalytic metal, the Lewis base, and O3′. And, it is likely that the conserved water at Mg_{cat} is involved, perhaps by its induced ionization to lose a proton to a water network with almost simultaneous transfer of the nearby just-added Asp256 proton. It is important to appreciate that there are at least three separated timescales in action here: the widely separated timescales of electrons and heavy nuclei (C,O,N), and the intermediate timescale of protons. The timescale of charge reorganization should be on the order of femtoseconds since it is electronic in nature. It is possible that the proton on the Lewis base at the Asp256 position (Pol β), once transferred from O3′, is short lived (it may be transferred through a water network that includes the conserved water molecule on the catalytic metal ion to the region of the NTP near Oαβ). All of these reasonable events will happen with the heavy nuclei essentially static. Instantaneously, we have the Pα–Oαβ bond weakened and the O3′–Pα interaction strengthened. As a result, inversion at Pα begins. Bond forming/breaking happens, perhaps nearly spontaneously. In this view, there is little activation energy of these steps; presumably the rate-controlling step that defines the turnover number is, in fact, the collective energy needed to snap all of the molecular parts in place (or perhaps to clear the active site once the reaction in over). Once in place—with the NTP symmetrically bridging the two metals at Pα and the O3′–Asp256 position hydrogen bond with both O3′ and the residue at the 256 position bound to the catalytic metal—the reaction is initiated by the tunneling event between O3′ and Asp256. (In the QM/MM simulations on Pol β (Batra et al., 2013; Lin et al., 2008, 2006), the proton was transferred to Asp256 in the slower process of classical barrier crossing and then retained on the Asp256 position Lewis base, while the insertion reaction was forced to occur. Tunneling of the proton was not evaluated and a sufficient water network to bridge Mg_{cat} and Oαβ was not in place.) The structural data on the DNA/RNA polymerases (Table 1) thus lead us to a view of polymerase reactivity that may only be fully revealed through the consideration of proton tunneling and multiple timescale events.

7. CONCLUSIONS

All QM/MM studies to date on polymerase reactions, either begin with models that do not correctly account for the necessary strong hydrogen bond at the O3′–Asp/Glu interaction, do not have enough water molecules present to allow for the proton to end up at O$\alpha\beta$ with little energy cost or have quantum subsystems which are charged, rather than neutral. The ideal precalculation crystal structure would have this key hydrogen bond in place, which requires the use of a nonhydrolyzable NTP. In this inferred ideal structure, the two magnesium ions would be about 3.5 Å apart, the NTP would be in place, and binding the Pα subunit symmetrically to the two metals and two aspartates would bridge the two magnesium ions. The model extracted from this would require three neutralizing positively charged amino acids around the periphery. These would ideally be arginines, which are less resistant to donating protons to the NTP than lysine. The structure of Pol β has three arginine groups ideally located for model building. Tunneling has generally not been considered in the DNA/RNA insertion chemistry reactions to date. Hopefully, future work in this area will strive to be described in such a manner as to be reasonably reproducible.

ACKNOWLEDGMENTS

This research is supported by Research Project Numbers Z01-ES050158 and Z01-ES050161 to S. H. W. and Z01-ES043010 to L. P. by the Intramural Research Program of the National Institutes of Health, National Institute of Environmental Health Sciences. L. G. P. acknowledges NIH Grant HL-06350.

REFERENCES

Abashkin, Y. G., Erickson, J. W., & Burt, S. K. (2001). Quantum chemical investigation of enzymatic activity of DNA polymerase β. A mechanistic study. *The Journal of Physical Chemistry B*, *105*, 287–292.

Ahn, J., Kraynov, V. S., Zhong, X., Werneburg, B. G., & Tsai, M. D. (1998). DNA polymerase β: Effects of gapped DNA substrates on dNTP specificity, fidelity, processivity and conformational changes. *The Biochemical Journal*, *331*, 79–87.

Bakhtina, M., Lee, S., Wang, Y., Dunlap, C., Lamarche, B., & Tsai, M. D. (2005). Use of viscogens, dNT PαS, and rhodium (III) as probes in stopped-flow experiments to obtain new evidence for the mechanism of catalysis by DNA polymerase β. *Biochemistry*, *55*, 5177–5187.

Bash, P. A., Field, M. J., Davenport, R. C., Petsko, G. A., Ringe, D., & Karplus, M. (1991). Computer simulation and analysis of the reaction pathway of triosephosphate isomerase. *Biochemistry, 30*, 5826–5832.

Batra, V. K., Beard, W. A., Shock, D. D., Krahn, J. M., Pedersen, L. C., & Wilson, S. H. (2006). Magnesium-induced assembly of a complete DNA polymerase catalytic complex. *Structure, 14*, 757–766.

Batra, V. K., Beard, W. A., Shock, D. A., Pedersen, L. C., & Wilson, S. H. (2008). Structure of DNA polymerase β with active-site mismatches suggest a transient abasic site intermediate during misincorporation. *Molecular Cell, 30*, 315–324.

Batra, V. K., Perera, L., Lin, P., Shock, D. D., Beard, W. A., Pedersen, L. C., et al. (2013). Amino acid substitution in the active site of DNA polymerase β explains the energy barrier of the nucleotidyl transfer reaction. *Journal of the American Chemical Society, 135*, 8078–8088.

Beard, W. A., & Wilson, S. H. (2014). Structure and mechanism of DNA polymerase. *Biochemistry, 53*, 2768–2780.

Becke, A. D. (1993). Density-functional thermochemistry. III. The role of exact exchange. *The Journal of Chemical Physics, 98*, 5648–5652.

Berg, B. L. V., Beard, W. L., & Wilson, S. H. (2001). DNA structure and Aspartate 276 influence nucleotide binding in human DNA polymerase beta. *Journal of Biological Chemistry, 276*, 3408–3416.

Besler, B. H., Merz, K. M., & Kollman, P. A. (1990). Atomic charges derived from semi-empirical methods. *Journal of Computational Chemistry, 4*, 431–439.

Biertumpfel, C., Zhao, Y., Kondo, Y., Ramon-Maiques, S., Gregory, M., Lee, J. Y., et al. (2010). Structure and mechanism of human DNA polymerase eta. *Nature, 465*, 1044–1048.

Brooks, B. R., Brooks, C. L., Mackerell, A. D., Nilsson, I., Petrella, R. J., Roux, B., et al. (2009). CHARMM: The biomolecular simulation package. *Journal of Computational Chemistry, 30*, 1545–1614.

Carvalho, A. T. P., Fernandes, P. A., & Ramos, M. J. (2011). The catalytic mechanism of RNA polymerase II. *Journal of Chemical Theory and Computation, 7*, 1177–1188.

Case, D. A., Cheatham, T. E., Darden, T., Gohlke, H., Luo, R., Merz, K. M., et al. (2005). The Amber biomolecular simulation programs. *Journal of Computational Chemistry, 26*, 1668–1688.

Case, D. A., Darden, T. A., Cheatham, T. E., Simmerling, C. L., Wang, J., Duke, R. E., et al. (2010). *Amber 11*. San Francisco: University of California.

Chung, L. W., Hirao, H., Li, X., & Morokuma, K. (2012). The ONIOM method: Its foundation and applications to metalloenzymes and photobiology. *Wiley Interdisciplinary Reviews. Computational Molecular Science, 2*, 327–350.

Cisneros, G. A., Perera, L., Garcia-Diaz, M., Bebenek, K., Kunkel, T. A., & Pedersen, L. G. (2008). Catalytic mechanism of human DNA polymerase lambda with Mg2+ and Mn2+ from ab initio quantum mechanical/molecular mechanical studies. *DNA repair, 7*, 1824–1834.

Darden, T., York, D., & Pedersen, L. (1993). Particle mesh Ewald: An N ln(N) method for Ewald sums in large systems. *The Journal of Chemical Physics, 98*, 10089–10092.

Ding, L., Chung, L. W., & Morokuma, K. (2013). Reaction mechanism of photoinduced decarboxylation of the photoactivatable free fluorescent protein: An ONIOM(QM: MM) study. *The Journal of Physical Chemistry. B, 117*, 1075–1084.

Essmann, Perera, L., Berkowitz, M. L., Darden, T., Lee, H., & Pedersen, L. G. (1995). A smooth particle mesh Ewald method. *The Journal of Chemical Physics, 103*, 8577–8593.

Foloppe, N., & Mackerell, A. D. (2000). All-atom empirical force field for nucleic acids. I. Parameter optimization based on small molecule and condensed phase macromolecular target data. *Journal of Computational Chemistry, 21*, 86–104.

Frisch, M. J., Trucks, G. W., Schlegel, H. B., Scuseria, G. E., Robb, M. A., Cheeseman, J. A., et al. (2004). *Gaussian 03, Revision D.02*. Wallingford, CT: Gaussian, Inc.

Garcia-Diaz, M., Bebenek, K., Krahn, J. M., Pedersen, L. C., & Kunkel, T. A. (2007). Role of the catalytic metal during polymerization by DNA polymerase λ. *DNA Repair, 6*, 1333–1340.

Groenhof, G. (2013). Introduction to QM/MM simulation. In L. Monticelli, & E. Salonen (Eds.), *Methods in molecular biology biomolecular simulations: Methods and protocols: Vol. 924* (pp. 43–66). New York, NY: Springer Science+Business Media.

Hay, S., & Scrutton, N. S. (2012). Good vibrations in enzyme-catalyzed reactions. *Nature Chemistry, 4*, 161–168.

Hu, L., Soderhjelm, P., & Ryde, U. (2011). On the convergence of QM/MM energies. *Journal of Chemical Theory and Computation, 7*, 761–777.

Hu, H., & Yang, W. (2008). Free energies of chemical reactions in solution and in enzymes with ab initio QM/MM methods. *Annual Review of Physical Chemistry, 59*, 573–601.

Klinman, J. P., & Kohen, A. (2013). Hydrogen tunneling links protein dynamics to enzyme catalysis. *Annual Review of Biochemistry, 82*, 471–496.

Klippenstein, S. J., Pande, V. S., & Truhlar, D. G. (2014). Chemical kinetics and mechanisms in complex systems: A perspective on recent theoretical advances. *Journal of the American Chemical Society, 136*, 538–546.

Konig, G., Hudson, P. S., Boresch, S., & Woodcock, H. L. (2014). Multiscale free energy simulations: An efficient method for connecting classical MD simulations to QM or QM/MM free energies using non-Boltzmann Bennett re-weighting schemes. *Journal of Chemical Theory and Computation, 10*, 1406–1419.

Kuznetsov, A. M., & Ulstrup, J. (1999). Proton and hydrogen atom tunneling in hydrolytic and redox enzyme catalysis. *Canadian Journal of Chemistry, 77*, 1085–1096.

Lee, C., Yang, W., & Parr, R. G. (1988). Development of the Colle-Salvetti correlation formula into a functional of the electron density. *Physical Review B, 37*, 785–789.

Lin, P., Batra, V. K., Pedersen, L. C., Beard, W. A., Wilson, S. H., & Pedersen, L. G. (2008). Incorrect nucleotide insertion at the active site of a G:A mismatch catalyzed by DNA polymerase β. *Proceedings of the National Academy of Sciences of the United States of America, 105*, 5670–5674.

Lin, P., Pedersen, L. C., Batra, V. K., Beard, W. A., Wilson, S. H., & Pedersen, L. G. (2006). Energy analysis of chemistry for correct insertion by DNA polymerase β. *Proceedings of the National Academy of Sciences of the United States of America, 103*, 13294–13299.

Lin, H., & Truhlar, D. G. (2007). QM/MM: What have we learned, where are we, and where do we go from here? *Theoretical Chemistry Accounts, 117*, 185–199.

Lior-Hoffmann, L., Wang, L., Wang, S., Geacintov, N. E., Broyde, S., & Zhang, Y. (2012). Preferred WMSA catalytic mechanism of the nucleotidyl transfer reaction in human DNA polymerase κ elucidates error-free bypass of a bulky lesion. *Nucleic Acids Research, 40*, 9193–9205.

Liu, X., Bushnell, D. A., Silva, D. A., Huang, X., & Kornberg, R. D. (2011). Initiation complex structure and promoter proofreading. *Science, 333*, 633–637.

Lone, S., Townson, S. A., Uljon, A. N., Johnson, R. E., Brahma, A., Nair, D. T., et al. (2007). Human DNA polymerase kappa encircles DNA: Implications for mismatch extension and lesion bypass. *Molecular Cell, 25*, 601–614.

MacKerrell, A. D., Bashford, D., Bellott, M., Dunbrack, R. L., Evanseck, J. D., Field, M. J., et al. (1998). All-atom empirical potential for molecular modeling and dynamics studies of proteins. *The Journal of Physical Chemistry. B, 102*, 3586–3616.

Moon, A. F., Proyr, J. M., Ramsden, D. A., Kunkel, T. A., Bebenek, K., & Pedersen, L. C. (2014). Sustained active site rigidity during synthesis by human DNA Polymerase μ. *Nature Structural and Molecular Biology, 21*, 253–260.

Mullholland, A. J., Roitberg, A. E., & Tunon, I. (2012). Enzyme and catalysis in the mechanism of DNA polymerase. *Theoretical Chemistry Accounts, 131*, 1286–1288.
Nam, K., Cui, Q., Gao, J., & York, D. M. (2007). Specific reaction parameterization of the AM1/d Hamiltonian for phosphoryl transfer reactions: H, O and P atoms. *Journal of Chemical Theory and Computation, 3*, 486–504.
Olson, A. C., Patro, J. N., Urban, M., & Kuchta, R. D. (2013). The energetic difference between synthesis of correct and incorrect base pairs account for highly accurate DNA replication. *Journal of the American Chemical Society, 135*, 1205–1208.
Phillips, J. C., Braun, R., Wang, W., Gumbart, J., Tajhorshid, E., Villa, E., et al. (2005). Scalable molecular dynamics with NAMD. *Journal of Computational Chemistry, 26*, 1781–1802.
Ponder, J. (1998). *TINKER. Software tools for molecular design, version 3.6*. St. Louis: Washington University.
Prasad, B. R., Kamerin, S. C. L., Florian, J., & Warshel, A. (2012). Prechemistry barriers and checkpoints do not contribute to fidelity and catalysis as long as they are not rate limiting. *Theoretical Chemistry Accounts, 131*, 1288–1303.
Radhakrishnan, R., & Schlick, T. (2004). Orchestration of cooperative events in DNA synthesis and repair mechanism unraveled by transition path sampling of DNA polymerase β's closing. *Proceedings of the National Academy of Sciences of the United States of America, 101*, 5970–5975.
Radhakrishnan, R., & Schlick, T. (2005). Fidelity discrimination in DNA polymerase β: Differing closing profiles for a mismatched (G:A) versus matched (G:C) base pair. *Journal of the American Chemical Society, 127*, 13245–13252.
Radhakrishnan, R., & Schlick, T. (2006). Correct and incorrect nucleotide incorporation pathways in DNA polymerase β. *Biochemical and Biophysical Research Communications, 350*, 521–529.
Rechkoblit, O., Malinina, L., Cheng, Y., Kuryavyi, V., Broyde, S., Geacintov, N. E., et al. (2006). Stepwise translocation of Dpo4 polymerase during error free bypass of an oxoG lesion. *PLoS Biology, 4*, e11.
Rittenhouse, R. C., Apostoluk, W. K., Miller, J. H., & Straatsma, T. P. (2003). Characterization of the active site of DNA polymerase β by molecular dynamics and quantum chemical calculations. *Proteins: Structure, Function, and Genetics, 53*, 667–682.
Rod, T. H., & Ryde, U. (2005). Quantum mechanical free energy barrier in enzymatic reaction. *Physical Review Letters, 94*, 138302.
Rosta, E., Yang, W., & Hummer, G. (2014). Calcium inhibition of ribonuclease H1 two-metal ion catalysis. *Journal of the American Chemical Society, 136*, 3137–3144.
Sawaya, M. R., Prasad, R., Wilson, S. H., Kraut, J., & Pelletier, H. (1997). Crystal structures of human DNA polymerase β complexed with gapped and nicked DNA: Evidence for an induced fit mechanism. *Biochemistry, 36*, 11205–11215.
Schlick, T., Arora, K., & Beard, W. A. (2012). Perspective: Pre-chemistry conformational changes in DNA polymerase reaction. *Theoretical Chemistry Accounts, 131*, 1287–1294.
Schmidt, M. M., Baldridge, K. K., Boatz, J. A., Elbert, S. T., Matsunaga, N., Nguyen, K. A., et al. (1993). General atomic and molecular electronic system. *Journal of Computational Chemistry, 15*, 1347–1363.
Scott, W. R. P., Hunenberger, P. H., Tironi, I. G., Mark, A. E., Billeter, S. R., Fennen, J., et al. (1999). The GROMOS biomolecular simulation package. *The Journal of Physical Chemistry. A, 103*, 3596–3607.
Senn, H. M., & Thiel, W. (2009). QM/MM methods for biomolecular systems. *Angewandte Chemie International Edition (English), 48*, 1198–1229.
Sobol, R. W. (2014). DNA repair polymerases. In K. S. Murakami, & M. A. Trakselsis (Eds.), *Nucleic acid polymerases: Vol. 30* (pp. 43–84). Berlin: Springer-Verlag.
Stewart, J. J. P. (1989). Optimization of parameters for semi-empirical methods. II. *Journal of Computational Chemistry, 10*, 221–264.

Truhlar, D. G. (2010). Tunneling in enzymatic and nonenzymatic hydrogen transfer reactions. *Journal of Physical Organic Chemistry, 23*, 660–676.

Truhlar, D. G., Gao, J., Garcia-Viloca, M., Alhambra, C., Corchado, J., Sanchez, M. L., et al. (2004). Ensemble-averaged variational transition state theory with optimized multidimensional tunneling for enzyme kinetics and other condenses-phase reactions. *International Journal of Quantum Chemistry, 100*, 1136–1152.

Tsai, M.-D. (2014). How DNA polymerases catalyze DNA replication, repair and mutation. *Biochemistry, 53*, 2749–2751.

Van der Kamp, M. W., & Mulholland, A. J. (2013). Combined quantum mechanics/molecular mechanics (QM/MM) methods in computational enzymology. *Biochemistry, 52*, 2708–2728.

Vreven, T., Byun, K. S., Komaromi, I., Dapprich, S., Montgomery, J. A., Morokuma, K., et al. (2006). Combining quantum mechanics methods with molecular mechanics methods in ONIOM. *Journal of Chemical Theory and Computation, 2*, 815–826.

Wadt, W. R., & Hay, P. J. (1985). Ab initio effective core potential for molecular calculation—Potentials for main group elements Na to Bi. *The Journal of Chemical Physics, 82*, 284–298.

Wang, D., Bushnell, D. A., Westover, K. D., Kaplan, C. D., & Kornberg, R. D. (2006). Structural basis of transcription role of the trigger loop in substrate specificity and catalysis. *Cell, 127*, 941–954.

Wang, Y., & Schlick, T. (2008). Quantum mechanics/molecular mechanics investigation of the chemical reaction in Dpo4 reveals water-dependent pathways and requirements for active site recognition. *Journal of the American Chemical Society, 130*, 13240–13250.

Wang, J., Wolf, R. M., Caldwell, J. W., Kollman, P. A., & Case, D. A. (2004). Development and testing of a general amber force field. *Journal of Computational Chemistry, 25*, 1157–1174.

Warshell, A., & Levitt, M. (1976). Theoretical studies of enzymatic reactions: Dielectric, electrostatic and steric stabilization of the carbonium ion in the reaction of lysozyme. *Journal of Molecular Biology, 103*, 227–249.

Wu, S., Beard, W. A., Pedersen, L. G., & Wilson, S. H. (2014). Structural comparison of DNA polymerase architecture suggests a nucleotide gate to the polymerase active site. *Chemical Reviews, 114*, 2759–2774.

Xia, S., Wang, M., Blaha, G., Konigsberg, W. H., & Wang, J. (2011). Structural insights into complete metal ion coordination from ternary complexes of a B family RB69 DNA polymerase. *Biochemistry, 50*, 9114–9224.

Zhang, Y. (2005). Improved pseudobonds for combined ab initio quantum mechanical/molecular mechanical (QM/MM) methods. *The Journal of Chemical Physics, 122*, 24224.

Zhang, R. (2013). *Multiscale simulation of mRNA synthesis by RNA polymerase II*. PhD thesis, University of Calgary.

Zhang, Y., Lee, T., & Yang, W. (1999). A pseudo-bond approach to combining quantum mechanical and molecular mechanical methods. *The Journal of Chemical Physics, 110*, 46–54.

Zhang, Y., Liu, H., & Yang, W. (2000). Free energy calculations on enzyme reactions with an efficient iterative procedure to determine minimum energy paths on a combined ab initio QM/MM potential energy surface. *The Journal of Chemical Physics, 112*, 3483–3492.

CHAPTER FIVE

Monitoring the Biomolecular Interactions and the Activity of Zn-Containing Enzymes Involved in Conformational Diseases: Experimental Methods for Therapeutic Purposes

Giuseppe Grasso[1]

Dipartimento di Scienze Chimiche, Università degli Studi di Catania, Catania, Italy
[1]Corresponding author: e-mail address: grassog@unict.it

Contents

1. Zn-Metalloproteases and Conformational Diseases — 116
2. Analytical Techniques Used to Study ZnMPs-Substrate/Inhibitors Interactions — 120
    2.1 Studying ZnMPs at atomic level — 121
    2.2 Analytical techniques able to provide information on binding affinities, activities, and substrate-induced conformational changes of ZnMPs — 125
    2.3 Mass spectrometry and ZnMPs — 131
3. Conclusions and Future Perspectives — 135
References — 135

Abstract

Zinc metalloproteases (ZnMPs) participate in diverse biological reactions, encompassing the synthesis and degradation of all the major metabolites in living organisms. In particular, ZnMPs have been recognized to play a very important role in controlling the concentration level of several peptides and/or proteins whose homeostasis has to be finely regulated for the correct physiology of cells. Dyshomeostasis of aggregation-prone proteins causes pathological conditions and the development of several different diseases. For this reason, in recent years, many analytical approaches have been applied for studying the interaction between ZnMPs and their substrates/inhibitors and how environmental factors can affect enzyme activities. In this scenario, nuclear magnetic resonance, X-ray diffraction, mass spectrometric (MS), and optical methods occupy a very important role in elucidating different aspects of the ZnMPs-substrates/inhibitors interaction, ranging from identification of cleavage sites to quantitation of kinetic parameters and inhibition constants. Here, an overview of all the main achievements in the application of different experimental approaches with special attention to MS methods to the investigation of ZnMPs-substrates/inhibitors interaction is given. A general MS experimental protocol which has been proved to be useful to study such interactions is also described.

1. ZN-METALLOPROTEASES AND CONFORMATIONAL DISEASES

Prior to the beginning of life, about 4×10^9 years ago, the atmosphere was reducing because of the dominant universal abundance of hydrogen over oxygen. In those conditions, zinc was estimated to be available in the sea at a concentration of less than 10^{-12} M and therefore life would not be possible on earth as we know it today (Williams & Fraústo Da Silva, 2004). Indeed, such low availability of zinc would hinder the use of this metal ion in the biological systems. However, once the level of oxygen was increased in the atmosphere, the oxidation of the elements was unavoidable and, as sulfide was oxidized to sulfate in the sea, zinc, as well as copper, became much more available to be used by the living systems as they reached concentrations up to 10^{-8} M and 10^{-10} M, respectively. Nowadays, zinc is present in and is essential to all forms of life. In higher mammals, zinc is the most abundant trace metal after iron. Particularly, it is widely present in more than 300 distinct enzymes, having structural and catalytic functions (Hooper, 1996). These Zn-metalloproteases (ZnMPs) comprise a vast family of proteins which are involved in many physiological processes fundamental for the life of cells and, more generally, of living organisms. The main reason for such a success of zinc is probably due to its redox stability, i.e., zinc is stable in the +2 oxidation state and it is neither oxidized or reduced in biological reactions. This property, together with its dimension and electronegativity, makes the Zn^{2+} ion particularly efficient as a polarizer of a water molecule inside the general ZnMP catalytic chamber (Grasso, Giuffrida, & Rizzarelli, 2012). Indeed, in the latter, zinc is maintained in position by coordinating residues (commonly cysteine, histidine, tyrosine, aspartic and glutamic acids) but it is also bound to water, thus generating a hydroxide ion that can attack the protein substrate (polarization-assisted zinc water catalysis) or induce the formation of nucleophiles (Vallee & Auld, 1990). Therefore, ZnMPs participate in biological reactions encompassing the degradation of all major metabolites (carbohydrates, lipids, nucleic acids, and proteins/peptides), and they can be classified according to their location (Taylor & Nicholson, 2003) or main function

(Crichton, 2008). In any case, ZnMPs are mostly responsible for the degradation and the catabolism of aggregation-prone biomolecules whose accumulation and fibrillogenesis seem to be crucial factors responsible for the development of conformational diseases such as Alzheimer's diseases (AD), Parkinson's disease, or prion diseases. Protein misfolding is an intrinsic aspect of normal folding within the complex cellular environment, and its effects are minimized in living systems by the action of a range of protective mechanisms that include molecular chaperones and quality control systems. ZnMPs play a very important role in regulating and controlling protein misfolding. Indeed, in physiological conditions ZnMPs contribute to lower the level of aggregation-prone proteins but several environmental factors can also contribute to protein misfolding such as interaction with metal ions (Loaiza et al., 2011; Tiiman, Palumaa, & Tougu, 2013), interaction with small molecules (Waermlaender et al., 2013), pH changes (Li, Xu, Mu, & Zhang, 2013), etc. (Bellotti et al., 2007). Amyloid fibrils and their precursors (oligomers) appear to have adverse effects on cellular functions regardless of the sequence of the component peptide or protein. Because of ZnMPs capability of modulating aggregation-prone protein levels, another method to sort ZnMPs would be to identify which ones are involved with the degradation of a particular substrate protein and, consequently, with a particular disease (Grasso & Bonnet, 2014). Table 1 lists the most widespread conformational diseases that have been associated with the aggregation of a specific protein (Dobson, 2001; Grasso & Spoto, 2013), together with the ZnMPs which are known to be involved either directly in the catabolism of the aggregation-prone protein, or indirectly with the pathological state. It is important to note that while some ZnMPs seem to be specific for a particular disease, many others are involved in different diseases. In many cases, this is due to the capability of some ZnMPs to degrade various protein substrates (Shen, Joachimiak, Rosner, & Tang, 2006). Indeed, it is practically impossible to group univocally the various existing ZnMPs and to associate them with a single disease, as most ZnMPs are involved in the development of several different pathologies. In this perspective, matrix MPs (MMPs) certainly represent the stereotype of the multifunctional ZnMP. They are zinc-dependent endopeptidases which belong to a larger family of proteases known as the metzincin superfamily. Collectively, they are capable of degrading all kinds of extracellular matrix proteins, but also can process a number of bioactive molecules and play an important role in tissue remodeling, which is itself associated with various physiological and pathological processes such as angiogenesis (Chen et al., 2014), morphogenesis

Table 1 Aggregation-prone proteins and ZnMPs involved with various conformational diseases

Conformational disease	Protein	ZnMPs involved	References
Hypercholesterolemia, atherosclerosis	Low-density lipoprotein receptor	MMP-12	Subramanian et al. (2012)
Cystic fibrosis	Cystic fibrosis trans-membrane regulator	MMP-9, MMP-13	Bensman, Nguyen, Rao, and Beringer (2012), Qu, Strickland, and Thomas (1997) and Nkyimbeng et al. (2013)
Huntington's disease	Huntingtin	MMP-9	Duran-Vilaregut et al., 2011
Marfan syndrome	Fibrillin	MMP-2, ADAMTSL6β	Xiong, Meisinger, Knispel, Worth, and Baxter (2012) and Saito et al. (2011)
Osteogenesis imperfecta	Procollagen	BMP1	Martínez-Glez et al. (2012)
Sickle cell anemia	Hemoglobin	MMP-2, MMP-9	Lee et al. (2007)
Scurvy	Collagen	MMP-1	Nusgens et al. (2001)
Alzheimer's disease	β-Amyloid, presenilin	NEP, IDE, ECE-1, ECE-2, ACE, MMPs, PreP, the proteasome	Malgieri and Grasso (2014)
Parkinson's disease	α-Synuclein, neuromelanin, lactoferrin, ferritin, ceruloplasmin, bivalent cation transporters	MMPs, ADAMs	Malgieri and Grasso (2014), Rosenberg (2009) and Gupta, Singh, Garg, Pant, and Khattri (2014)
Scrapie/Creutzfeldt–Jakob disease	Prion protein, ferritin	ADAM8, ADAM9, ADAM10	Malgieri and Grasso (2014), Altmeppen et al. (2012), Taylor et al. (2009) and Liang et al. (2012)

Table 1 Aggregation-prone proteins and ZnMPs involved with various conformational diseases—cont'd

Conformational disease	Protein	ZnMPs involved	References
Familial amyloidoses	Transthyretin/lysozyme	MMP-9	Almeida and Saraiva (2012)
Retinitis pigmentosa	Rhodopsin	ADAM9	Parry et al. (2009)
Cataracts	Crystallins	MMPs, ADAM, ADAMTS	Robertson, Siwakoti, and West-Mays (2013) and Hodgkinson, Wang, Duncan, Edwards, and Wormstone (2010)
Cancer	p53	MMPs, IDE, ADAMs	Tundo et al. (2013) and Moro, Mauch, and Zigrino (2014)
Friedreich's ataxia	Frataxin, aconitase	MPP	Gordon, Shi, Dancis, and Pain (1999)
Progressive supranuclear palsy	α-Synuclein	MMP-1, MMP-9	Lorenzl, Albers, Chirichigno, Augood, and Beal (2004)
Wilson's disease	Ceruloplasmin, Wilson's protein	MMP-2, MMP-12	Sokolov et al. (2009)
Type II diabetes	Insulin, amylin	IDE, NEP	Bellia and Grasso (2014) and Oefner, Pierau, Schulz, and Dale (2007)
Carotid atherosclerosis	Proteins in vessel walls	MMP-9, MMP-12	Silvello et al. (2014) and Scholtes et al. (2012)
Lewy-body dementia	α-Synuclein	MMPs	Lorenzl, Buerger, Hampel, and Beal (2008)
Familial amyotrophic lateral sclerosis	Superoxide dismutase 1	MMP-3, MMP-9	Dewil et al. (2005) and Lee et al. (2008)

Readapted from Grasso and Bonnet (2014).

(Detry et al., 2012), tissue repair (Gajendrareddy et al., 2013), cirrhosis (Liu et al., 2013), systemic sclerosis (Peng et al., 2012), and metastasis (Moss, Jensen-Taubman, & Stetler-Stevenson, 2012). Recent data suggested an active role of MMPs in the pathogenesis of aortic aneurysm and a dysregulation of the balance between MMPs and their natural inhibitors (TIMPs) is also a characteristic of acute and chronic cardiovascular diseases (Raffetto & Khalil, 2008). This multiplicity of functions makes the design of drugs targeting specific ZnMPs a very appealing therapeutic approach, as well as a challenging task (Grasso & Bonnet, 2014).

In this perspective, the scientific community has recently focused on the design of specific ZnMPs modulators which should be able to influence the catabolism of aggregation-prone proteins and, consequently, have a therapeutic value for tackling conformational diseases. In order to do so, it is imperative to have as much information as possible about the structure of the ZnMP under study (for best modulator design) as well as of the ZnMP-substrate/inhibitor complex (for substrate specific inhibition). Moreover, in order to establish modulators performances and specificity, it is also necessary to monitor ZnMPs activity in the presence of the various biomolecular partners and at the different environmental conditions which have a biological significance *in vivo*. For these purposes, many experimental approaches have been applied on various ZnMPs-substrate/inhibitor systems and an overview of those is given in the next sections.

2. ANALYTICAL TECHNIQUES USED TO STUDY ZnMPs-SUBSTRATE/INHIBITORS INTERACTIONS

The interaction between ZnMPs and their substrates and/or modulators can be studied by many different experimental approaches which can give information on different aspects of such interaction. Particularly, we can distinguish between the analytical approaches that give an insight on the structure of the ZnMP-substrate/modulator adduct and the ones that give information on binding affinities, activity, and substrate/inhibitor-induced conformational changes of the enzymes without elucidating molecular details and/or binding sites. In sections 2.1 and 2.2 following paragraphs, the main analytical approaches used in both cases will be discussed. Finally, in section 2.3 the potentiality of mass spectrometry (MS) to study different aspects of the ZnMPs investigation is also presented.

2.1. Studying ZnMPs at atomic level

Nowadays, various analytical techniques (i.e., XAS, IR, etc.) are available to the experimentalists which are interested to study, at atomic level, ZnMPs interaction with substrates and/or activity modulators. However, in this section, only the two most successfully used ones will be discussed.

2.1.1 Nuclear magnetic resonance

Nuclear magnetic resonance (NMR) spectroscopy is one of the most valuable tool to identify and characterize binding sites in proteins. It can provide, at an atomic resolution level, information about the structure and dynamics of a protein in solution, allowing then to map the structural and dynamic changes that take place upon substrate binding (Bertini et al., 2012; Xu et al., 2009). In order to monitor such changes, it is necessary to collect a large number of information separately, choosing each time the appropriate pulse sequence and recording different spectra (Cavanagh, Fairbrother, Palmer, & Skelton, 1996). Advances in NMR instrumentation and methods have greatly facilitated this task. Commonly, the proteins are isotopically labeled with ^{13}C and ^{15}N (either selectively or uniformly) and a large number of sequences (multidimensional either 2, 3, or 4D experiments) are nowadays available in literature. In large proteins, however, the magnetization relaxes faster, causing the peaks to become weaker and eventually disappear. Spin–spin relaxation can be reduced by working with samples in which non-exchangeable protons have been fractionally or fully deuterated, an approach which leads to amazing improvements in spectral quality. The TROSY (Pervushin, Riek, Wider, & Wüthrich, 1997) version of the experiments is often used that takes advantage of the partial cancelation of the dipolar and CSA (chemical shift anisotropy) relaxation mechanisms to generate sharp resonances for molecules whose weights are over 100 kDa. In some cases, small oligopeptides mimicking the catalytic chamber of large ZnMPs have been synthesized in order to overcome the above mentioned problems (Arus et al., 2013) or to answer to specific questions regarding the coordination of the Zn ion inside the catalytic chamber of different classes of ZnMPs (Grasso et al., 2014). Analogously, in order to understand the coordination chemistry of Zn-binding groups with catalytic Zn centers in MMPs and ADAMs, NMR has also been successfully applied (He, Puerta, Cohen, & Rodgers, 2005).

Another important and common application of NMR is the investigation of the mechanisms of folding of ZnMPs and the way such mechanisms are influenced by the presence of a substrate. For copper-containing enzyme, paramagnetism-assisted NMR (Fragai, Luchinat, Parigi, & Ravera, 2013) is an extremely useful tool for the characterization of conformational disorder in proteins and protein complexes: the paramagnetism-based restraints are in fact very sensitive to the nuclear-paramagnetic center distances and to the relative orientations of nuclear-paramagnetic center and nuclear–nuclear vectors. Therefore, they can provide information on the conformations that can be mostly populated by looking at the local changes at atomic resolution. In the case of ZnMPs, the binding of zinc, a diamagnetic metal ion, to proteins is usually indirectly studied through the "chemical shift mapping" method (Jensen, Hass, Hansen, & Led, 2007). In the latter approach, the NMR spectra are examined to record the perturbations of the chemical shifts in titration experiments where the concentration of the zinc ion is progressively increased. Such perturbations, due to the structural changes caused by the metal binding and to the electric field induced by its charge, allow to localize the metal-binding site (Malgieri & Grasso, 2014). Although such method does not provide information about the structure of the metal-binding sites, the relative position of the coordinating residues and the surrounding structure of the protein close to the metal ion can be obtained from the distances (<5 Å) derived from the Nuclear Overhauser Enhancements (NOEs). In some cases, the changes in ^{19}F NMR signal from ^{19}F-Trp residues can be also used to indicate structural changes which would be no detectable by other experimental approaches such as X-ray studies (Niu et al., 2010). Unfortunately, the only NMR-active isotope of zinc, ^{67}Zn, has a natural abundance of only 4.11% and is characterized by a large quadrupole moment (nuclear spin $I=5/2$) and high relaxation rates. For these reasons, ^{67}Zn NMR has been profitably used to characterize Zn^{2+}-protein-binding features only in few cases (Shimizu & Hatano, 1985).

The use of triple-resonance 3D NMR experiments has also been exploited in order to map the structural and dynamic changes that take place upon substrate and/or inhibitor binding to ZnMPs. For example, assignments for the main-chain NMR chemical shifts and delineation of the secondary structure of the catalytic domain of human stromelysin-1 complexed with a hydrophobic inhibitor has been obtained from such approach. It has been therefore possible to assess that the inhibitor binds in an extended conformation in the hydrophobic pockets of the active site (Van Doren et al., 1995).

In order to identify the Zn-binding amino acidic residues, it is important to note that histidine ligands can be recognized through $^1\text{H}-^{15}\text{N}$ HSQC spectra optimized for the detection of 2J(N,H) couplings (Baglivo et al., 2009). For the identification of other Zn-binding amino acidic residues such as cysteines, one can rely on the chemical shift of ^{13}C nuclei (C_α and C_β for cysteines), which are generally very sensitive to the presence of a bond to a metal ion (Baglivo et al., 2009). There is also the possibility to substitute the Zn ion with another NMR-active metal such as ^{113}Cd and ^{199}Hg, so that the metal–ligand couplings can be detected by heteronuclear 2D experiments such as metal–proton correlation spectra. This method can be used not only to identify the nature of the protein ligands in uncharacterized cases but also to unveil the dynamics at the metal-binding site (Armitage, Drakenberg, & Reilly, 2013).

Finally, in addition to the structural changes just mentioned, NMR spectroscopy can provide detailed information about the flexibility of ZnMPs, especially when is combined with X-ray studies. As an example, full-length MMP-12 has been demonstrated to have relative mobility of its catalytic and hemopexin (HPX) domains by observing that the R_1 and R_2 (relaxation times) values are intermediate between those of the isolated domains and those expected for any rigid structure of the full-length protein (Bertini et al., 2008). Indeed, the solution structure of the HPX domain of MMP-12 was solved from the NMR signals and the full-length protein was assigned. It was so possible to obtain the relaxation data (R_1, R_2, NOE) for the full-length protein and compared it with the same data for its isolated catalytic and HPX domains. These data showed that the two domains are not held rigidly to one another but must undergo independent motions.

2.1.2 X-ray crystallography

Over the past decade, X-ray crystallography has been often coupled with NMR for the investigation of ZnMPs, mostly due to the recent developments in biotechnology, instrumentation and computational methods (Isaksson et al., 2009). Both techniques share the ability to determine time and spatial averages of molecular properties at atomic level. For both techniques, average properties are determined from measurements performed on an ensemble of molecules and it is possible, within such ensemble, to have information about variability on the type and relative population of conformational substates. However, beyond the above mentioned similarities, significant differences between NMR and X-ray crystallography are evident, especially regarding the spatial distribution of the molecules in the sample and time scales accessible to each method. Indeed, NMR data represent averages in the time range of nanosecond to second; on the other side, diffraction data represent averages in the time range of second to hours. The difference in the spatial distribution investigated by NMR and X-ray diffraction is also significant by considering that while the former gives information from an average over randomly oriented molecules, the latter samples molecules arranged in a periodic crystal lattice. In this respect, solid-state NMR is the most similar NMR approach to X-ray crystallography, but in this case the analysis can be expanded to microcrystalline and amorphous materials which are amenable to this method (Grasso & Titman, 2009) (interactions of very short range) but not to X-ray, which cannot be applied in case of a lack of single crystals suitable for diffraction (Macholl, Lentz, Boerner, & Buntkowsky, 2007).

Despite these limitations, X-ray crystallography has been widely applied to study ZnMPs structures and interactions with substrates as it overcomes some limitations encountered in NMR studies, such as its problematic applicability on very large systems. Moreover, the limits imposed by the crystallization can be in some cases overcome by recurring to the soaking approach. For example, in the case of poor co-crystallization of a ZnMP with an inhibitor, the soaking approach has been successfully applied (Belviso et al., 2013), allowing to obtain a snapshot of the first reaction step of the inhibition process of $[PtCl_3(DMSO)]^-$ on MMP-3. In this case, Pt binds to His224 of MMP-3, before any further rearrangement can occur. This event, possibly favored by crystal packing, brings the Met143 and Glu139 functional groups close to one another, while in solution the enzyme folding leaves great conformational freedom to the amino acid side chains, thus explaining why coordination to this site is not observed in the solution experiment (Arnesano et al., 2009).

Strikingly, X-ray crystallography has been also applied to monitor the reaction mechanism of some ZnMPs (Bertini et al., 2006). This has been achieved by obtaining a series of X-ray crystal structures of different MMPs (MMP-8 and MMP-12) under a variety of conditions to obtain models of the various steps of the reaction mechanism of such MMPs. Surprisingly, although the authors were not able to identify a peptide bound form of the enzymes because of rapid hydrolysis (the peptide used was ProGlnGlyIleAlaGly, which is known to be cleaved at the Gly–Ile bond by MMPs), well-resolved X-ray crystallographic structures showing a hydrolysis product inside the catalytic chamber for both MMP-12 and MMP-8 could be obtained. These structures could then be used to build a plausible model for the series of events that likely take place in the catalytic cycle of MMPs. The same group was also able to obtain extremely high resolution of structures of the catalytic domain of MMP-12 in the presence of two different inhibitors (acetohydroxamic acid and N-isobutyl-N-[4-methoxyphenylsulfonyl]glycyl hydroxamic acid) (Bertini et al., 2005). These structures, together with the one having batimastat as the inhibitor (Lang et al., 2001), which has been previously obtained by another group, have been compared and, due to the very high resolution (1.0–1.3 Å), it was possible to interpret any small deviation (>0.5 Å) among atomic coordinates of different molecules in terms of molecular flexibility. NMR investigations in the solution state was also carried out and the major finding was that there are certain loop regions that are subject to mobility and/or conformational heterogeneity in several MMPs. This research clearly shows once again the advantages of comparing X-ray structures with the solution ones obtained, for example, by NMR, in order to obtain a structural model that is beyond a single structural datum.

2.2. Analytical techniques able to provide information on binding affinities, activities, and substrate-induced conformational changes of ZnMPs

Although many analytical techniques have been applied to obtain information on binding affinities, activities, and substrate/inhibitor-induced conformational changes of ZnMPs, in the following only some particularly advantageous and most diffused optical methods will be discussed.

2.2.1 Surface plasmon resonance

Surface plasmon resonance (SPR) is a label-free analytical technique which is able to measure the binding between target analyte molecules and receptors previously immobilized onto a gold surface (Grasso & Spoto, 2013). During the receptor/analyte binding event, the shift of the dip in the spectrum of reflected light from the gold surface is monitored over time and kinetics information regarding biomolecules interactions are obtained (Grasso, Bush, D'Agata, Rizzarelli, & Spoto, 2009). Advantages of the SPR approach are the short time required for the analysis (few minutes in most cases) and the little sample consumption (in the nanomole–picomole range). However, this techniques does not give molecular information regarding the interaction, so that, in order to obtain a clear picture of the biomolecular events, the coupling with other analytical techniques is often needed (Arnold et al., 2011). SPR has been widely applied to investigate the interaction between MMPs and their substrates/inhibitors and the most common approach is to immobilize the substrate/inhibitor onto the gold surface and study the interaction with the enzyme in the flowing solution (Olson, Gervasi, Mobashery, & Fridman, 1997; Shoji, Kabeya, & Sugawara, 2011). This approach has been preferred in the case of MMPs because these enzymes experience autohydrolysis once they are immobilized on the surface, hindering the positive outcome of the SPR investigation. However, some authors have nonetheless performed such investigation by immobilizing the ZnMPs and applying a crosslinking procedure for stabilization of the surface. In this way, the unfolding of the enzyme that is required for autohydrolysis was hindered and meaningful results could be obtained as long as the drift in the SPR signal was taken into account (Gossas et al., 2013). In this way, it was found that different inhibitors have very different dissociation phases and those containing a hydroxamate moiety were discovered to dissociate very slowly from the enzyme–inhibitor complex (see Fig. 1). Although determination of the mechanism and kinetic parameters of slowly dissociating inhibitors had proven to be challenging also with such biosensor-based approach, it allowed a semi-quantitative analysis that is very informative for lead discovery.

In any case, it is important to bear in mind that in order to obtain meaningful results by SPR, comparative experiments need to be performed. Therefore, for example, only the comparison of the SPR obtained kinetic parameters for the interaction between MMP-9 and native triple-helical collagen IV on one side and its catalytic domain with the same substrate on the other side allowed to reveal that the association constant of MMP-9 is much larger than that of its catalytic domain. Such result strongly suggests that the interplay among hemopexin-like domain, fibronectin type II repeats motif, and linker region (O-glycosylated domain) plays an important role in recognizing collagen IV (Shoji et al., 2011). In this case as in many others, molecular information could not be obtained from absolute SPR kinetic values, but by performing parallel experiments on similar systems differing from each other only for the variable which has to be investigated.

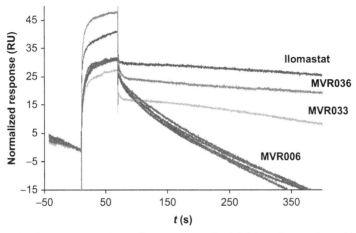

Figure 1 Qualitative comparison of sensorgrams for inhibitors interacting with MMP-12. Sensorgrams of 3, 4, and 5 were obtained with different enzyme surfaces, while one sensorgram of 1 was obtained for each surface and used to normalize the binding capacities of the different surfaces. All inhibitors were injected at a concentration of 5 uM. The sensorgrams are solvent corrected, adjusted for differences in molecular weight of inhibitors and blank subtracted. *Reprinted with permission from Gossas et al. (2013).*

2.2.2 Other optical methods

If it is true that the understanding of the biomolecular mechanisms involved in the interaction between ZnMPs and their substrates/inhibitors is crucial for the rational design of new therapeutic approaches, it has to be highlighted that the possibility to monitor enzyme activity *in vitro* and *in vivo* also represents a necessary step toward the development of possible therapeutic drugs based on enzyme activity modulation. Indeed, once the important role that ZnMPs have on several physiological and pathological processes has been established and recognized, new demands for methods aimed at rapid and reliable evaluation of enzymatic activities as well as chemical screening of enzyme inhibitors have risen (Reymond, 2006). Many attempts based on the optical sensing have been made to analyze the activity of ZnMPs, but most of them require an expensive fluorescence readout and a labor-intensive process. As ZnMPs activity is finely tuned *in vivo* by complex mechanisms, including spatial and temporal expression, small-molecule binding and posttranslational modifications, the search for methods to monitor enzyme activity *in vivo* has been specifically pursued (Zhu et al., 2014). Indeed, it became obvious that it would be possible to image tumors *in vivo* by means of tumor-associated proteolytic activity (Scherer, McIntyre, & Matrisian, 2008). These imaging approaches target MMPs involved in cancer progression via contrast agents linked to MMP inhibitors or to MMP selective and specific substrates with sensitivity enhanced by amplification during enzymatic processing. Particularly, near-infrared (NIR) light (650–900 nm) is especially suitable for noninvasive *in vivo* imaging, because it is relatively poorly absorbed by biomolecules and can penetrate well through tissues (Myochin, Hanaoka, Komatsu, Terai, & Nagano, 2012). Moreover, background autofluorescence is low in the NIR region, making the use of NIR fluorogenic probes highly attractive. Recently, tailored designed activatable NIR fluorescent probes with signal amplification properties for the *in vivo* detection of cancer-related MMP activity have been described (see Fig. 2) (Akers et al., 2012).

An alternative approach to image ZnMPs activity without a fluorescent label relies on the use of an activatable quantum dot (QD) for cancer detection (Zheng et al., 2007). Compared to conventional fluorophores, QDs offer brighter signal, better photo-stability, and the potential for integration into multifunctional reagents. Moreover, colorimetric assays using gold nanoparticles (AuNPs) have also been applied for clinical and environmental

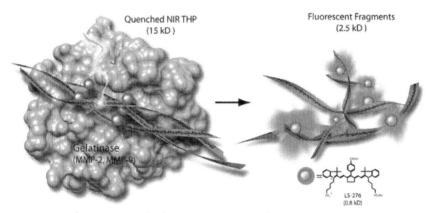

Figure 2 NIR fluorescent probe for *in vivo* detection of cancer-related MMP activity. The probe is based on a triple-helical peptide substrate (THP) with high specificity for MMP-2 and MMP-9 relative to other members of the MMP family. At the center of each 5-kDa peptide strand is a MMP-sensitive sequence flanked by two Lys residues conjugated with NIR fluorescent dyes. Upon self-assembly of the triple-helical structure, the three peptide chains intertwine, bringing the fluorophores into close proximity and reducing fluorescence via quenching (left). Upon enzymatic cleavage of the triple-helical peptide, six labeled peptide chains are released, resulting in an amplified fluorescent signal (right). *Reprinted with permission from Akers et al. (2012).*

diagnoses due to their simplicity and low cost. Self-assembly of AuNPs typically brings about a color change, which is attributed to particle–particle plasmonic coupling and scattering effects. AuNPs are therefore used in this case because of their distinct optical properties, such as high extinction coefficients and localized SPR (Grasso & Spoto, 2013). As an example, an ADAMTS-4 detective fluorescent turn-on AuNP probe (ADAMTS-4-D-Au probe) has been developed by conjugating AuNPs with a FITC-modified ADAMTS-4-specific peptide (DVQEFRGVTAVIR) (Peng et al., 2013). The authors demonstrated that this probe can be used to detect ADAMTS-4 in biological samples, representing a potential biomarker for early diagnosis of cartilage-damage diseases.

Additionally, it is important to mention that in order to detect substrate/inhibitor-induced conformational changes of ZnMPs, circular dichroism (CD) has been widely applied. CD is normally used to obtain information about the secondary structure of the ZnMP as the spectrum of the peptide bands (around 190–230 nm) depends on the content of α-helix, β-sheet, turns, and random coil. Typically, the far-UV CD spectra of polypeptides with extensive α-helical structures have two characteristic minima near 208 and 222 nm. β-Sheet structure yields a minimum at about 215 nm and random coil is characterized by lack of a positive peak at about 195 nm, a negative peak in the vicinity of 200 nm, and low ellipticity at about 222 nm. For example, in the case of vibriolysin, a ZnMP present in the *Vibrio cholera*, the far-UV CD spectrum shows a major negative dichroic absorption band at 218 nm, which is characteristic of proteins containing both α-helices and β-sheets (Iqbal, Azim, Hashmi, Ali, & Musharaf, 2011). In this case, the authors concluded that mature vibriolysin contains an $\alpha+\beta$ protein fold and the α-helices are the predominant secondary structural components.

More interestingly, changes in the conformational structure induced by biomolecules that can act as substrate/activity modulators of ZnMPs can be easily monitored by CD. Recently, modulation of insulin-degrading enzyme (IDE) activity has attracted much interest as a therapeutic target for various diseases, from diabetes to AD (Grasso et al., 2011; Grasso, Rizzarelli, & Spoto, 2008, 2009; Grasso, Salomone, et al., 2012). Particularly, somatostatin has been found to be both a substrate and a modulator of IDE (Ciaccio et al., 2009) and it was possible to monitor the conformational change of the enzyme upon somatostatin addition to the solution. In Fig. 3, the recorded CD spectra are reported together with the somatostatin concentration-dependent effect on the ellipticity signal of active and EDTA inhibited IDE at 207 nm. From the fitting of the data, it was possible to draw

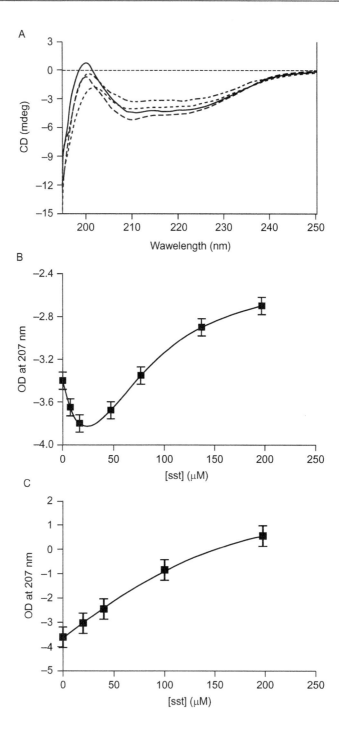

the conclusion that somatostatin binds IDE by two different sites with different affinities, one of them being the catalytic site (somatostatin is degraded by IDE). By inhibiting the enzyme with EDTA, the affinity value for the high-affinity site for somatostatin (possibly the active site) is dramatically reduced (Fig. 3C).

Finally, CD can also be applied to investigate the changes in the secondary structure and/or in the oligomeric state of a ZnMP upon different environmental conditions. For example, Yi, Gruszczynska-Biegala, Wood, Zhao, and Zolkiewska (2005) compared the CD spectrum of the extracellular domain of ADAM12 (amino acids 32–707) (1) with the sum (2) of the two CD spectra obtained from the separated (i) prodomain and metalloprotease domain (amino acids 32–421); and (ii) the disintegrin, cysteine-rich, and EGF-like domains and the recombinant proteins (amino acids 422–706). A large deviation between (1) and (2) was observed and the authors could conclude that the secondary structures of the autonomously expressed prodomain metalloprotease domain and the disintegrin/cysteine-rich/EGF-like domains differ from the structures present in the intact extracellular domain of ADAM12. Although this result does not exclude the possibility that the two separated domains both participate in forming a protein-binding site required for cell adhesion, it strongly suggests that the autonomously expressed fragments do not assume the same secondary structures that are present in full extracellular domain of ADAM12.

2.3. Mass spectrometry and ZnMPs

The possibility to investigate biomolecules by MS has been widely increased by the availability of soft ionization techniques such as electrospray (ESI) and matrix-assisted laser induced ionization (MALDI). In both cases, the detection of the whole enzymes as well as of the enzyme–inhibitor complexes are feasible at convenient low concentrations. In addition to the simple detection of the complex species, it is also possible to identify the binding sites

Figure 3 Conformational changes in IDE structure mediated by somatostatin. (A) CD spectroscopy of 1 µM IDE in 40 mM phosphate buffer at 37 °C in the absence and in the presence of different somatostatin concentrations: 0 µM (—-), 40 µM (– – –), 80 µM (- - -), and 200 µM (–··–··–). (B) The somatostatin concentration-dependent effect on the ellipticity signal of IDE at 207 nm. Continuous line was obtained from the nonlinear least-squares fitting of data (see Ciaccio et al., 2009). (C) The somatostatin effect on IDE inactivated with EDTA at 207 nm. The continuous line corresponds to a single binding curve. *Reprinted with permission from Ciaccio et al. (2009).*

within a ZnMP by recurring to the hydrogen/deuterium exchange (HDX) of protein backbone with detection by MS (Garcia et al., 2005). Briefly, a ZnMP or a ZnMP/inhibitor pair is incubated for defined intervals in a deuterated environment. After rapid quenching of the HDX reaction, the partially deuterated protein is digested, and the resulting peptide fragments are analyzed by liquid chromatography (LC)-MS. It is then possible to correlate the deuterium buildup curve measured for each peptide fragment with its environment in the intact protein. Consequently, the peptidic parts of the ZnMP that experience the most different amounts of HDX in the presence and in the absence of the inhibitor must be the regions of the ZnMP which are either mainly involved with the binding and/or undergo the largest inhibitor-induced conformational change. In this way, three noncatalytic regions in MMP-1 (residues 285–295, 302–316, and 437–457) and even two specific residues (Ile-290 and Arg-291) have been identified to participate in collagen proteolysis (Lauer-Fields et al., 2009). Particularly, Ile-290 and Arg-291 seem to contribute to the recognition of the triple-helical structure and facilitate both the binding and catalysis of the triple helix. It was so possible to infer that the MMP-1 catalytic and hemopexin-like domains collaborate in collagen catabolism by properly aligning the triple helix and coupling conformational states to facilitate hydrolysis. Besides giving information on possible ZnMPs binding sites, the use of more technological advanced MS approaches such as traveling wave ion mobility spectrometry-MS have also allowed to perform a speciation analysis of ZnMPs under different experimental conditions. For example, a study of the influence of pH and time of incubation on the complex species formed by different metal ions with carbonic anhydrase has been reported (de Souza Pessôa, Pilau, Gozzo, & Zezzi Arruda, 2013). It was so possible to identify several complexed species having different mobilities from the apo form of this protein.

From the few examples discussed above, it is clear that MS can be listed among the useful analytical techniques which are able to provide the answers to the questions raised by scientists interested in ZnMPs. However, it is important to highlight that one of the main advantages of MS is its versatility. Indeed, besides identification and characterization of the ZnMPs complexes with substrates/inhibitors and binding sites, MS has its larger applications in determining both the ZnMPs activity and the associated amount and types of peptide fragments produced by ZnMPs proteolytic activity (Bellia & Grasso, 2014; Bellia, Pietropaolo, & Grasso, 2013; Grasso, Bush, et al., 2009; Grasso et al., 2005, 2008; Grasso, Fragai, Rizzarelli, Spoto, & Yeo, 2007; Grasso,

Rizzarelli, et al., 2009; Prely et al., 2012). For example, it is straightforward to identify the cleavage sites observed by MALDI time of flight-MS experiments after the action of the catalytic site of MMP-3 on α-synuclein (Sung et al., 2005). In this case, the quantitative analysis of the peptide fragments formed at different incubation times allowed to establish that α-synuclein is degraded by MMP-3 from its C-terminal end. This experimental approach is general and is normally applied to identify cleavage sites preferences in peptides degradation by ZnMPs. But why is it so important to assess cleavage site preferences and to identify ZnMPs produced peptide fragments? In recent years, it has become clear that peptide fragments which are formed from degradation of a biological molecule can have biochemical functions which largely differ from their precursors (Autelitano et al., 2006). In many cases, it is even very difficult to predict the biological activities of the fragments either from their amino acid sequence or the activity of the precursor protein. An example is given by the fragment 9–23 of the B chain of insulin which has been suggested to be a primary autoantigenic epitope in the pathogenesis of type 1 diabetes in NOD mice (Moriyama et al., 2003; Nakayama et al., 2005). Moreover, immunization of the latter with exogenous B9-23 peptide has been reported to prevent diabetes (Daniel & Wegmann, 1996). However, the biochemical mechanisms through which such insulin fragment is generated *in vivo* are unknown, while some *in vitro* evidences have been recently reported (Bellia et al., 2013). The analytical approach applied in this case consists of performing an enzymatic digestion of a peptide by a ZnMP and analyzing all the generated fragments by LC-MS. In this way, it is possible not only to identify the cleavage sites of the ZnMPs from the fragmentation but also to quantify the relative amounts of the produced peptide fragments at different experimental conditions. For example, the binding of metal ions to the peptide substrate can hinder the proteolysis by the ZnMP only to certain cleavage sites which are close to the metal-binding site and not to others (Bellia & Grasso, 2014; Bellia et al., 2013). As an example, in Fig. 4, the bar graphs relative to all the chromatographic areas ascribed to any peptide detected by HPLC-MS, formed by the IDE proteolytic action on (A) insulin A chain and (B) insulin B chain, with and without zinc(II) for several incubation times (0–9 h) are reported and it seems that, through IDE proteolytic activity, the zinc(II) homeostasis could regulate the production of some specific insulin fragments (fragment 9–23 of the B chain among the others). The change in cleavage sites preferentiality at the different experimental conditions (metal ions binding in this case) experienced by the peptide substrate is only detectable by this MS method, as all the other

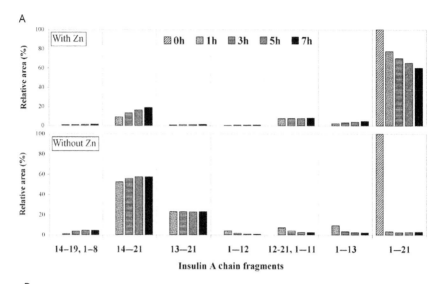
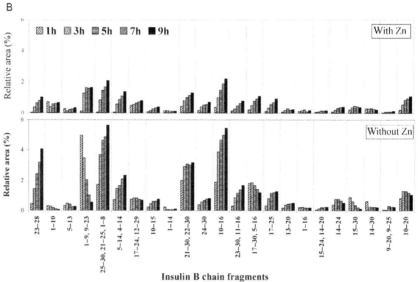

Figure 4 Bar graphs relative to all the chromatographic areas ascribed to any peptide detected by HPLC-MS, formed by the IDE proteolytic action on (A) insulin A chain and (B) insulin B chain, with and without zinc(II) at 37 °C for several incubation times (0–9 h). The chromatographic area of all peaks are reported as relative to the total area of all the detected peaks of each chromatogram. *Reprinted with permission from Bellia et al. (2013).*

analytical tools monitoring enzyme activity can only detect the overall activity changes. However, in the light of the above discussed different activities that the various biopeptides originating from the same precursor protein can have *in vivo*, the knowledge of the relative abundances of the generated peptide fragments can be of paramount importance. In this perspective, this LC-MS combined approach is a very useful and general tool for progressing in the knowledge of ZnMPs, as several proteins of this class of enzymes are capable of generating many not yet well-characterized bioactive peptides ("cryptic" peptides) and the mechanism by which the latter are formed has to be still clarified in most cases, adding complexity to the overall challenge of the so-called cryptome (Pimenta & Lebrun, 2007).

3. CONCLUSIONS AND FUTURE PERSPECTIVES

From the discussion carried out above, I hope it became clear that, in my opinion, the best way to proceed when confronted with a scientific problem concerning ZnMPs (but surely this is valid for any topic) is to look at it from different perspectives, that is by different analytical approaches. Indeed, although it is very useful to being able to monitor ZnMPs activities at the various experimental conditions, at the same time it is necessary to being able to assess the molecular mechanisms which are responsible for the enzyme activity modulation. For this reason, besides all the optical analytical techniques which are able to give information on the overall activity changes, other structural techniques such as NMR and X-ray diffraction should be always employed to give an insight at atomic level. In this scenario, MS-based approaches, alone or coupled with other techniques, can be of great help in assessing the action of ZnMPs at specific substrate cleavage sites. A great deal of collaboration between the various ZnMPs labs with different expertise is therefore very much desirable for future progress in the field.

REFERENCES

Akers, W. J., Xu, B., Lee, H., Sudlow, G. P., Fields, G. B., Achilefu, S., et al. (2012). Detection of MMP-2 and MMP-9 activity in vivo with a triple-helical peptide optical probe. *Bioconjugate Chemistry, 23*, 656–663.

Almeida, M. R., & Saraiva, M. J. (2012). Clearance of extracellular misfolded proteins in systemic amyloidosis: Experience with transthyretin. *FEBS Letters, 586*, 2891–2896.

Altmeppen, H. C., Puig, B., Dohler, F., Thurm, D. K., Falker, C., Krasemann, S., et al. (2012). Proteolytic processing of the prion protein in health and disease. *American Journal of Neurodegenerative Disease, 1*, 15–31.

Armitage, I. M., Drakenberg, T., & Reilly, B. (2013). Use of (113)Cd NMR to probe the native metal binding sites in metalloproteins: An overview. *Metal Ions in Life Sciences*, *11*, 117–144.

Arnesano, F., Boccarelli, A., Cornacchia, D., Nushi, F., Sasanelli, R., Coluccia, M., et al. (2009). Mechanistic insight into the inhibition of matrix metalloproteinases by platinum substrates. *Journal of Medicinal Chemistry*, *52*, 7847–7855.

Arnold, L. H., Butt, L. E., Prior, S. H., Read, C. M., Fields, G. B., & Pickford, A. R. (2011). The interface between catalytic and hemopexin domains in matrix metalloproteinase-1 conceals collagen binding exosite. *Journal of Biological Chemistry*, *286*, 45073–45082.

Arus, D., Nagy, N. V., Dancs, A., Jancso, A., Berkecz, R., & Gajda, T. (2013). A minimalist chemical model of matrix metalloproteinases—Can small peptides mimic the more rigid metal binding sites of proteins? *Journal of Inorganic Biochemistry*, *126*, 61–69.

Autelitano, D. J., Rajic, A., Smith, A. I., Berndt, M. C., Ilag, L. L., & Vadas, M. (2006). The cryptome: A subset of the proteome, comprising cryptic peptides with distinct bioactivities. *Drug Discovery Today*, *11*, 306–314.

Baglivo, I., Russo, L., Esposito, S., Malgieri, G., Renda, M., Salluzzo, A., et al. (2009). The structural role of the zinc ion can be dispensable in prokaryotic zinc-finger domains. *Proceedings of the National Academy of Sciences of the United States of America*, *106*, 6933–6938.

Bellia, F., & Grasso, G. (2014). The role of copper(II) and zinc(II) in the degradation of human and murine IAPP by insulin-degrading enzyme. *Journal of Mass Spectrometry*, *49*, 274–279.

Bellia, F., Pietropaolo, A., & Grasso, G. (2013). Formation of insulin fragments by insulin-degrading enzyme: The role of zinc(II) and cystine bridges. *Journal of Mass Spectrometry*, *48*, 135–140.

Bellotti, V., Nuvolone, M., Giorgetti, S., Obici, L., Palladini, G., Russo, P., et al. (2007). The workings of the amyloid diseases. *Annals of Medicine*, *39*, 200–207.

Belviso, B. D., Caliandro, R., Siliqi, D., Calderone, V., Arnesano, F., & Natile, G. (2013). Structure of matrix metalloproteinase-3 with a platinum-based inhibitor. *Chemical Communications*, *49*, 5492–5494.

Bensman, T. J., Nguyen, A. N., Rao, A. P., & Beringer, M. P. (2012). Doxycycline exhibits anti-inflammatory activity in CF bronchial epithelial cells. *Pulmonary Pharmacology and Therapeutics*, *25*, 377–382.

Bertini, I., Calderone, V., Cosenza, M., Fragai, M., Lee, Y.-M., Luchinat, C., et al. (2005). Conformational variability of matrix metalloproteinases: Beyond a single 3D structure. *Proceedings of the National Academy of Sciences of the United States of America*, *102*, 5334–5339.

Bertini, I., Calderone, V., Fragai, M., Jaiswal, R., Luchinat, C., Melikian, M., et al. (2008). Evidence of reciprocal reorientation of the catalytic and hemopexin-like domains of full-length MMP-12. *Journal of the American Chemical Society*, *130*, 7011–7021.

Bertini, I., Calderone, V., Fragai, M., Luchinat, C., Maletta, M., & Yeo, K. J. (2006). Snapshots of the reaction mechanism of matrix metalloproteinases. *Angewandte Chemie International Edition*, *45*, 7952–7955.

Bertini, I., Fragai, M., Luchinat, C., Melikian, M., Toccafondi, M., Lauer, J. L., et al. (2012). Structural basis for matrix metalloproteinase 1-catalyzed collagenolysis. *Journal of the American Chemical Society*, *134*, 2100–2110.

Cavanagh, J., Fairbrother, W. J., Palmer, A. G., III., & Skelton, N. J. (1996). *Protein NMR Spectroscopy. Principles and Practice*. San Diego: Academic Press.

Chen, Y., Huang, Y., Huang, Y., Xia, X., Zhang, J., Zhou, Y., et al. (2014). JWA suppresses tumor angiogenesis via Sp1-activated matrix metalloproteinase-2 and its prognostic significance in human gastric cancer. *Carcinogenesis*, *35*, 442–451.

Ciaccio, C., Tundo, G. R., Grasso, G., Spoto, G., Marasco, D., Ruvo, M., et al. (2009). Somatostatin: A novel substrate and a modulator of insulin degrading enzyme activity. *Journal of Molecular Biology*, *385*, 1556–1567.

Crichton, R. R. (2008). *Biological inorganic chemistry: An introduction.* Amsterdam: Elsevier.
Daniel, D., & Wegmann, D. R. (1996). Protection of nonobese diabetic mice from diabetes by intranasal or subcutaneous administration of insulin peptide B-(9–23). *Proceedings of the National Academy of Sciences of the United States of America, 93,* 956–960.
de Souza Pessôa, G., Pilau, E. J., Gozzo, F. C., & Zezzi Arruda, M. A. (2013). Ion mobility spectrometry focusing on speciation analysis of metals/metalloids bound to carbonic anhydrase. *Analytical and Bioanalytical Chemistry, 405,* 7653–7660.
Detry, B., Erpicum, C., Paupert, J., Blacher, S., Maillard, C., Bruyere, F., et al. (2012). Matrix metalloproteinase-2 governs lymphatic vessel formation as an interstitial collagenase. *Blood, 119,* 5048–5056.
Dewil, M., Schurmans, C., Starckx, S., Opdenakker, G., Van Den Bosch, L., & Robberecht, W. (2005). Role of matrix metalloproteinase-9 in a mouse model for amyotrophic lateral sclerosis. *NeuroReport, 16,* 321–324.
Dobson, C. M. (2001). The structural basis of protein folding and its links with human disease. *Philosophical Transactions of the Royal Society B, 356,* 133–145.
Duran-Vilaregut, J., del Valle, J., Manich, G., Camins, A., Pallas, M., Vilaplana, J., et al. (2011). Role of matrix metalloproteinase-9 (MMP-9) in striatal blood-brain barrier disruption in a 3-nitropropionic acid model of Huntington' disease. *Neuropathology and Applied Neurobiology, 37,* 525–537.
Fragai, M., Luchinat, C., Parigi, G., & Ravera, E. (2013). Conformational freedom of metalloproteins revealed by paramagnetism-assisted NMR. *Coordination Chemistry Reviews, 257,* 2652–2667.
Gajendrareddy, P. K., Engeland, C. G., Junges, R., Horan, M. P., Rojas, I. G., & Marucha, P. T. (2013). MMP-8 overexpression and persistence of neutrophils relate to stress-impaired healing and poor collagen architecture in mice. *Brain, Behavior, and Immunity, 28,* 44–48.
Garcia, R. A., Pantazatos, D. P., Gessner, C. R., Go, K. V., Woods, V. L., Jr., & Villarreal, F. J. (2005). Molecular interactions between matrilysin and the matrix metalloproteinase inhibitor doxycycline investigated by deuterium exchange mass spectrometry. *Molecular Pharmacology, 67,* 1128–1136.
Gordon, D. M., Shi, Q., Dancis, A., & Pain, D. (1999). Maturation of frataxin within mammalian and yeast mitochondria: One-step processing by matrix processing peptidase. *Human Molecular Genetics, 8,* 2255–2262.
Gossas, T., Nordstroem, H., Xu, M.-H., Sun, Z.-H., Lin, G.-Q., Wallberg, H., et al. (2013). The advantage of biosensor analysis over enzyme inhibition studies for slow dissociating inhibitors—Characterization of hydroxamate-based matrix metalloproteinase-12 inhibitors. *Medicinal Chemistry Communications, 4,* 432–442.
Grasso, G., & Bonnet, S. (2014). Metal complexes and metalloproteases: Targeting conformational diseases. *Metallomics, 6*(8), 1346–1357. http://dx.doi.org/10.1039/C4MT00076E.
Grasso, G., Bush, A. I., D'Agata, R., Rizzarelli, E., & Spoto, G. (2009). Enzyme solid-state support assays: A surface plasmon resonance and mass spectrometry coupled study of immobilized insulin degrading enzyme. *European Biophysics Journal, 38,* 407–414.
Grasso, G., D'Agata, R., Rizzarelli, E., Spoto, G., D'Andrea, L., Pedone, C., et al. (2005). Activity of anchored human matrix metalloproteinase-1 catalytic domain on Au (111) surfaces monitored by ESI-MS. *Journal of Mass Spectrometry, 40,* 1565–1571.
Grasso, G., Fragai, M., Rizzarelli, E., Spoto, G., & Yeo, K. J. (2007). A new methodology for monitoring the activity of cdMMP-12 anchored and freeze-dried on Au (111). *Journal of the American Society for Mass Spectrometry, 18,* 961–969.
Grasso, G., Giuffrida, M. L., & Rizzarelli, E. (2012). Metallostasis and amyloid β-degrading enzymes. *Metallomics, 4,* 937–949.
Grasso, G., Magrì, A., Bellia, F., Pietropaolo, A., La Mendola, D., & Rizzarelli, E. (2014). The copper(II) and zinc(II) coordination mode of HExxH and HxxEH motif in small

peptides: The role of carboxylate location and hydrogen bonding network. *Journal of Inorganic Biochemistry*, *130*, 92–102.

Grasso, G., Pietropaolo, A., Spoto, G., Pappalardo, G., Tundo, G. R., Ciaccio, C., et al. (2011). Copper(I) and copper(II) inhibit Aβ peptides proteolysis by insulin-degrading enzyme differently: Implications for metallostasis alteration in Alzheimer's disease. *Chemistry-A European Journal*, *17*, 2752–2762.

Grasso, G., Rizzarelli, E., & Spoto, G. (2008). How the binding and degrading capabilities of insulin degrading enzyme are affected by ubiquitin. *Biochimica et Biophysica Acta*, *1784*, 1122–1126.

Grasso, G., Rizzarelli, E., & Spoto, G. (2009). The proteolytic activity of insulin-degrading enzyme: A mass spectrometry study. *Journal of Mass Spectrometry*, *44*, 735–741.

Grasso, G., Salomone, F., Tundo, G. R., Pappalardo, G., Ciaccio, C., Spoto, G., et al. (2012). Metal ions affect insulin-degrading enzyme activity. *Journal of Inorganic Biochemistry*, *117*, 351–358.

Grasso, G., & Spoto, G. (2013). Plasmonics for the study of metal ion–protein interactions. *Analytical and Bioanalytical Chemistry*, *405*, 1833–1843.

Grasso, G., & Titman, J. J. (2009). Chain folding and diffusion in monodisperse long n-alkanes by solid-state NMR. *Macromolecules*, *42*, 4175–4180.

Gupta, V., Singh, M. K., Garg, R. K., Pant, K. K., & Khattri, S. (2014). Evaluation of peripheral matrix metalloproteinase-1 in Parkinson's disease: A case-control study. *International Journal of Neuroscience*, *124*, 88–92.

He, H., Puerta, D. T., Cohen, S. M., & Rodgers, K. R. (2005). Structural and spectroscopic study of reactions between chelating zinc-binding groups and mimics of the matrix metalloproteinase and disintegrin metalloprotease catalytic sites: The coordination chemistry of metalloprotease inhibition. *Inorganic Chemistry*, *44*, 7431–7442.

Hodgkinson, L. M., Wang, L., Duncan, G., Edwards, D. R., & Wormstone, I. M. (2010). ADAM and ADAMTS gene expression in native and wound healing human lens epithelial cells. *Molecular Vision*, *16*, 2765–2776.

Hooper, N. M. (1996). *Zinc metalloproteases in health and disease*. London: Taylor & Francis.

Iqbal, A., Azim, M. K., Hashmi, N., Ali, S. A., & Musharaf, S. G. (2011). Structural characterization of metalloprotease vibriolysin of cholera pathogen Vibrio cholerae. *Protein and Peptide Letters*, *18*, 287–294.

Isaksson, J., Nystroem, S., Derbyshire, D., Wallberg, H., Agback, T., Kovacs, H., et al. (2009). Does a fast nuclear magnetic resonance spectroscopy- and X-ray crystallography hybrid approach provide reliable structural information of ligand-protein complexes? A case study of metalloproteinases. *Journal of Medicinal Chemistry*, *52*, 1712–1722.

Jensen, R. M., Hass, M. A. S., Hansen, D. F., & Led, J. J. (2007). Investigating metal-binding in proteins by nuclear magnetic resonance. *Cellular and Molecular Life Sciences*, *64*, 1085–1104.

Lang, R., Kocourek, A., Braun, M., Tschesche, H., Huber, R., Bode, W., et al. (2001). Substrate specificity determinants of human macrophage elastase (MMP-12) based on the 1.1 Å crystal structure. *Journal of Molecular Biology*, *312*, 731–742.

Lauer-Fields, J. L., Chalmers, M. J., Busby, S. A., Minond, D., Griffin, P. R., & Fields, G. B. (2009). Identification of specific hemopexin-like domain residues that facilitate matrix metalloproteinase collagenolytic activity. *Journal of Biological Chemistry*, *284*, 24017–24024.

Lee, S. P., Ataga, K. I., Zayed, M., Manganello, J. M., Orringer, E. P., Phillips, D. R., et al. (2007). Phase I study of eptifibatide in patients with sickle cell anaemia. *British Journal of Haematology*, *139*, 612–620.

Lee, J. K., Shin, J. H., Suh, J. H., Choi, I. S., Ryu, K. S., & Gwag, B. J. (2008). Tissue inhibitor of metalloproteinases-3 (TIMP-3) expression is increased during serum deprivation-induced neuronal apoptosis in vitro and in the G93A mouse model of amyotrophic lateral sclerosis and tential modulator of Fas-mediated apoptosis. *Neurobiology of Disease*, *30*, 174–185.

Li, Y., Xu, W., Mu, Y., & Zhang, J. Z. H. (2013). Acidic pH retards the fibrillization of human islet amyloid polypeptide due to electrostatic repulsion of histidines. *Journal of Chemical Physics, 139*, 055102/1–055102/9.

Liang, J., Wang, W., Sorensen, D., Medina, S., Ilchenko, S., Kiselar, J., et al. (2012). Cellular prion protein regulates its own α-cleavage through ADAM8 in skeletal muscle. *Journal of Biological Chemistry, 287*, 16510–16520.

Liu, J., Cheng, X., Guo, Z., Wang, Z., Li, D., Kang, F., et al. (2013). Truncated active human matrix metalloproteinase-8 delivered by a chimeric adenovirus-hepatitis B virus vector ameliorates rat liver cirrhosis. *PLoS One, 8*, e53392.

Loaiza, A., Ronau, J. A., Ribbe, A., Stanciu, L., Burgner, J. W., 2nd., Pauland, L. N., et al. (2011). Folding dynamics of phenylalanine hydroxylase depends on the enzyme's metallation state: The native metal, iron, protects against aggregate intermediates. *European Biophysics Journal, 40*, 959–968.

Lorenzl, S., Albers, D. S., Chirichigno, J. W., Augood, S. J., & Beal, M. F. (2004). Elevated levels of matrix metalloproteinases-9 and -1 and of tissue inhibitors of MMPs, TIMP-1 and TIMP-2 in postmortem brain tissue of progressive supranuclear palsy. *Journal of the Neurological Sciences, 218*, 39–45.

Lorenzl, S., Buerger, K., Hampel, H., & Beal, M. F. (2008). Profiles of matrix metalloproteinases and their inhibitors in plasma of patients with dementia. *International Psychogeriatrics, 20*, 67–76.

Macholl, S., Lentz, D., Boerner, F., & Buntkowsky, G. (2007). Polymorphism of N,N"-diacetylbiuret studied by solid-state ^{13}C and ^{15}N NMR spectroscopy, DFT calculations, and X-ray diffraction. *Chemistry-A European Journal, 13*, 6139–6149.

Malgieri, G., & Grasso, G. (2014). The clearance of misfolded proteins in neurodegenerative diseases by zinc metalloproteases: An inorganic perspective. *Coordination Chemistry Reviews, 260*, 139–155.

Martínez-Glez, V., Valencia, M., Caparrós-Martín, J. A., Aglan, M., Temtamy, S., Tenorio, J., et al. (2012). Identification of a mutation causing deficient BMP1/mTLD proteolytic activity in autosomal recessive osteogenesis imperfecta. *Human Mutation, 33*, 343–350.

Moriyama, H., Abiru, N., Paronen, J., Sikora, K., Liu, E., Miao, D., et al. (2003). Evidence for a primary islet autoantigen (preproinsulin 1) for insulitis and diabetes in the nonobese diabetic mouse. *Proceedings of the National Academy of Sciences of the United States of America, 100*, 10376–10381.

Moro, N., Mauch, C., & Zigrino, P. (2014). Metalloproteinases in melanoma. *European Journal of Cell Biology, 93*, 23–29.

Moss, L. A. S., Jensen-Taubman, S., & Stetler-Stevenson, W. G. (2012). Matrix metalloproteinases: Changing roles in tumor progression and metastasis. *American Journal of Pathology, 181*, 1895–1899.

Myochin, T., Hanaoka, K., Komatsu, T., Terai, T., & Nagano, T. (2012). Design strategy for a near-infrared fluorescence probe for matrix metalloproteinase utilizing highly cell permeable boron dipyrromethene. *Journal of the American Chemical Society, 134*, 13730–13737.

Nakayama, M., Abiru, N., Moriyama, H., Babaya, N., Liu, E., Miao, D., et al. (2005). Prime role for an insulin epitope in the development of type 1 diabetes in NOD mice. *Nature, 435*, 220–223.

Niu, W., Shu, Q., Chen, Z., Mathews, S., Di Cera, E., & Frieden, C. (2010). The role of Zn^{2+} on the structure and stability of murine adenosine deaminase. *Journal of Physical Chemistry B, 114*, 16156–16165.

Nkyimbeng, T., Ruppert, C., Shiomi, T., Dahal, B., Lang, G., Seeger, W., et al. (2013). Pivotal role of matrix metalloproteinase 13 in extracellular matrix turnover in idiopathic pulmonary fibrosis. *PLoS One, 8*, e73279.

Nusgens, B. V., Humbert, P., Rougier, A., Colige, A. C., Haftek, M., Lambert, C. A., et al. (2001). Topically applied vitamin C enhances the mRNA level of collagens I and III, their processing enzymes and tissue inhibitor of matrix metalloproteinase 1 in the human dermis. *Journal of Investigative Dermatology, 116*, 853–859.

Oefner, C., Pierau, S., Schulz, H., & Dale, G. E. (2007). Structural studies of a bifunctional inhibitor of neprilysin and DPP-IV. *Acta Crystallographica Section D, 63*, 975–981.

Olson, M. W., Gervasi, D. C., Mobashery, S., & Fridman, R. (1997). Kinetic analysis of the binding of human matrix metalloproteinase-2 and -9 to tissue inhibitor of metalloproteinase (TIMP)-1 and TIMP-2. *Journal of Biological Chemistry, 272*, 29975–29983.

Parry, D. A., Toomes, C., Bida, L., Danciger, M., Towns, K. V., McKibbin, M., et al. (2009). Loss of the metalloprotease ADAM9 leads to cone-rod dystrophy in humans and retinal degeneration in mice. *American Journal of Human Genetics, 84*, 683–691.

Peng, W.-J., Yan, J.-W., Wan,, Y.-N., Wang, B.-X., Tao, J.-H., Yang, G.-J., et al. (2012). Matrix metalloproteinases: A review of their structure and role in systemic sclerosis. *Journal of Clinical Immunology, 32*, 1409–1414.

Peng, S., Zheng, Q., Zhang, X., Dai, L., Zhu, J., Pi, Y., et al. (2013). Detection of ADAMTS-4 activity using a fluorogenic peptide-conjugated Au nanoparticle probe in human knee synovial fluid. *ACS Applied Materials & Interfaces, 5*, 6089–6096.

Pervushin, K., Riek, R., Wider, G., & Wüthrich, K. (1997). Attenuated T2 relaxation by mutual cancellation of dipole-dipole coupling and chemical shift anisotropy indicates an avenue to NMR structures of very large biological macromolecules in solution. *Proceedings of the National Academy of Sciences of the United States of America, 94*, 12366–12371.

Pimenta, D. C., & Lebrun, I. (2007). Cryptides: Buried secrets in proteins. *Peptides, 28*, 2403–2410.

Prely, L. M., Paal, K., Hermans, J., van der Heide, S., van Oosterhout, A. J. M., & Bischoff, R. (2012). Quantification of matrix metalloprotease-9 in bronchoalveolar lavage fluid by selected reaction monitoring with microfluidics nano-liquid-chromatography-mass spectrometry. *Journal of Chromatography A, 1246*, 103–110.

Qu, B.-H., Strickland, E., & Thomas, P. J. (1997). Localization and suppression of a kinetic defect in cystic fibrosis transmembrane conductance regulator folding. *Journal of Biological Chemistry, 272*, 15739–15744.

Raffetto, J. D., & Khalil, R. A. (2008). Matrix metalloproteinases and their inhibitors in vascular remodeling and vascular disease. *Biochemical Pharmacology, 75*, 346–359.

Reymond, J. L. (Ed.), (2006). *Enzyme assays: High-throughput screening, genetic selection and fingerprinting*. New York: Wiley–VCH.

Robertson, J. V., Siwakoti, A., & West-Mays, J. A. (2013). Altered expression of transforming growth factor beta 1 and matrix metalloproteinase-9 results in elevated intraocular pressure in mice. *Molecular Vision, 19*, 684–695.

Rosenberg, G. A. (2009). Matrix metalloproteinases and their multiple roles in neurodegenerative diseases. *Lancet Neurology, 8*, 205–216.

Saito, M., Kurokawa, M., Oda, M., Oshima, M., Tsutsui, K., Kosaka, K., et al. (2011). ADAMTSL6β protein rescues fibrillin-1 microfibril disorder in a marfan syndrome mouse model through the promotion of fibrillin-1 assembly. *Journal of Biological Chemistry, 286*, 38602–38613 S38602/1–S38602/14..

Scherer, R. L., McIntyre, J. O., & Matrisian, L. M. (2008). Imaging matrix metalloproteinases in cancer. *Cancer Metastasis Review, 27*, 679–690.

Scholtes, V. P. W., Johnson, J. L., Jenkins, N., Sala-Newby, G. B., de Vries, J.-P. P. M., de Borst, G. J., et al. (2012). Carotid atherosclerotic plaque matrix metalloproteinase-12-positive macrophage subpopulation predicts adverse outcome after endarterectomy. *Journal of the American Heart Association, 1*, 001040/1–001040/13.

Shen, Y., Joachimiak, A., Rosner, M. R., & Tang, W.-J. (2006). Structures of human insulin-degrading enzyme reveal a new substrate recognition mechanism. *Nature, 443*, 870–874.

Shimizu, T., & Hatano, M. (1985). Magnetic resonance studies of trifluoperazine-calmodulin solutions: Calcium-43, magnesium-25, zinc-67, and potassium-39 nuclear magnetic resonance. *Inorganic Chemistry, 25*, 2003–2009.

Shoji, A., Kabeya, M., & Sugawara, M. (2011). Real-time monitoring of matrix metalloproteinase-9 collagenolytic activity with a surface plasmon resonance biosensor. *Analytical Biochemistry, 419*, 53–60.

Silvello, D., Narvaes, L. B., Albuquerque, L. C., Forgiarini, L. F., Meurer, L., Martinelli, N. C., et al. (2014). Serum levels and polymorphisms of matrix metalloproteinases (MMPs) in carotid artery atherosclerosis: Higher MMP-9 levels are associated with plaque vulnerability. *Biomarkers, 19*, 49–55.

Sokolov, A. V., Pulina, M. O., Ageeva, K. V., Tcherkalina, O. S., Zakharova, E. T., & Vasilyev, V. B. (2009). Identification of complexes formed by ceruloplasmin with matrix metalloproteinases 2 and 12. *Biochemistry, 74*, 1388–1392.

Subramanian, V., Uchida, H. A., Ijaz, T., Moorleghen, J. J., Howatt, D. A., & Balakrishnan, A. (2012). Calpain inhibition attenuates angiotensin II-induced abdominal aortic aneurysms and atherosclerosis in low-density lipoprotein receptor-deficient mice. *Journal of Cardiovascular Pharmacology, 59*, 66–76.

Sung, J. Y., Park, S. M., Lee, C. H., Um, J. W., Lee, H. J., Kim, J., et al. (2005). Proteolytic cleavage of extracellular secreted α-synuclein via matrix metalloproteinases. *Journal of Biological Chemistry, 280*, 25216–25224.

Taylor, K. M., & Nicholson, R. I. (2003). The LZT proteins; the LIV-1 subfamily of zinc transporters. *BBA-Biomembranes, 1611*, 16–30.

Taylor, D. R., Parkin, E. T., Cocklin, S. L., Ault, J. R., Ashcroft, A. E., Turner, A. J., et al. (2009). Role of ADAMs in the ectodomain shedding and conformational conversion of the prion protein. *Journal of Biological Chemistry, 284*, 22590–22600.

Tiiman, A., Palumaa, P., & Tougu, V. (2013). The missing link in the amyloid cascade of Alzheimer's disease—Metal ions. *Neurochemistry International, 62*, 367–378.

Tundo, G. R., Sbardella, D., Ciaccio, C., Bianculli, A., Orlandi, A., Desimio, M. G., et al. (2013). Insulin-degrading enzyme (IDE). *Journal of Biological Chemistry, 288*, 2281–2289.

Vallee, B. L., & Auld, D. S. (1990). Active-site zinc ligands and activated H2O of zinc enzymes. *Proceedings of the National Academy of Sciences of the United States of America, 87*, 220–224.

Van Doren, S. R., Kurochkin, A. V., Hu, W., Ye, Q. Z., Johnson, L. L., Hupe, D. J., et al. (1995). Solution structure of the catalytic domain of human stromelysin complexed with a hydrophobic inhibitor. *Protein Science, 4*, 2487–2498.

Waermlaender, S., Tiiman, A., Abelein, A., Luo, J., Jarvet, J., Soederberg, K. L., et al. (2013). Biophysical studies of the amyloid β-peptide: Interactions with metal ions and small molecules. *ChemBioChem, 14*, 1692–1704.

Williams, R. J. P., & Fraústo Da Silva, J. J. R. (2004). The trinity of life: The genome, the proteome and the mineral chemical elements. *Journal of Chemical Education, 81*, 738–749.

Xiong, W., Meisinger, T., Knispel, R., Worth, J. M., & Baxter, B. T. (2012). MMP-2 regulates erk1/2 phosphorylation and aortic dilatation in Marfan syndrome. *Circulation Research, 110*, e92–e101.

Xu, X., Mikhailova, M., Ilangovan, U., Chen, Z., Yu, A., Pal, S., et al. (2009). Nuclear magnetic resonance mapping and functional confirmation of the collagen binding sites of matrix metalloproteinase-2. *Biochemistry, 48*, 5822–5831.

Yi, H., Gruszczynska-Biegala, J., Wood, D., Zhao, Z., & Zolkiewska, A. (2005). Cooperation of the metalloprotease, disintegrin, and cysteine-rich domains of ADAM12 during inhibition of myogenic differentiation. *Journal of Biological Chemistry, 280*, 23475–23483.

Zheng, G., Chen, J., Stefflova, K., Jarvi, M., Li, H., & Wilson, B. C. (2007). Photodynamic molecular beacon as an activatable photosensitizer based on protease-controlled singlet oxygen quenching and activation. *Proceedings of the National Academy of Sciences of the United States of America, 21*, 8989–8994.

Zhu, L., Ma, Y., Kiesewetter, D. O., Wang, Y., Lang, L., Lee, S., et al. (2014). Rational design of matrix metalloproteinase-13 activatable probes for enhanced specificity. *ACS Chemical Biology, 9*, 510–516.

AUTHOR INDEX

Note: Page numbers followed by "*f*" indicate figures, "*t*" indicate tables and "*ge*" indicate glossary.

A

Abashkin, Y. G., 87–88, 88*f*, 89, 91–92
Abelein, A., 116–120
Abiru, N., 132–135
Abriata, L. A., 54
Acharya, K. R., 38–39, 39*f*
Achilefu, S., 127–128, 128*f*
Adams, M. W., 57–58
Addlagatta, A., 58–59
Adir, N., 5–6, 17, 28, 28*f*
Adolph, H.-W., 50–52, 53, 56–57
Agback, T., 123–124
Ageeva, K. V., 118*t*
Aglan, M., 118*t*
Agostinelli, E., 70–71
Ahn, H. J., 70–71
Ahn, J., 89–90, 94–95
Akers, W. J., 127–128, 128*f*
Alarcon, R., 71
Albers, D. S., 118*t*
Albuquerque, L. C., 118*t*
Alessi, C. M., 57–58
Alhambra, C., 107–109
Ali, S. A., 129
Allen, S. H., 65–66
Allona, I., 65–66
Almeida, M. R., 118*t*
Al-Oweini, R., 5–6, 28–29
Altmeppen, H. C., 118*t*
Altunkaya, A., 6
Alves, A., 53
Amicosante, G., 53
Amrhein, N., 65–66
Andersson, G., 65–66
Angel, N. Z., 69
Anne, C., 50–51, 53, 54
Antony, J., 51–52
Antonyuk, S. V., 65–66
Apostoluk, W. K., 89, 91–92
Aquino, M. A., 67–68
Arad-Yellin, R., 41

Aragoncillo, C., 65–66
Armitage, I. M., 123
Arnesano, F., 124
Arnold, L. H., 40, 41–43, 42*f*, 125–126
Arnold, W. N., 65–66
Arora, K., 97–98
Arus, D., 121
Arvas, M., 5, 7, 15
Asanuma, M., 8
Ash, D. E., 70–71
Ashcroft, A. E., 118*t*
Asiedu, E. T., 57–58
Asther, M., 6, 8
Ataga, K. I., 118*t*
Augood, S. J., 118*t*
Auld, D. S., 116–120
Ault, J. R., 118*t*
Autelitano, D. J., 132–135
Averill, B. A., 65–66, 68
Azim, M. K., 129

B

Babaya, N., 132–135
Badarau, A., 50–51, 54, 56–57, 70
Baglivo, I., 123
Bak, H. J., 3–5
Bakhtina, M., 94–95
Balakrishnan, A., 118*t*
Baldwin, M. J., 2–3, 17–18
Banovic, L., 53
Barbagallo, R. N., 6
Bash, P. A., 85
Bashford, D., 105–106
Bastiaan-Net, S., 7–8
Bateson, J. H., 54
Batista, S. C., 61–62, 65–66, 67–68
Batra, V. K., 87, 88*f*, 89–97, 91*f*, 92*f*, 98–101, 102–103, 107–109, 107*t*
Bauer, K., 57–58
Bauer, R., 51–52, 56–57
Baugh, L., 70–71

143

Baumbach, G. A., 65–66
Baxter, B. T., 118t
Bazer, F. W., 65–66
Beal, M. F., 118t
Beard, W. A., 83–114
Beate, J., 66–67
Bebenek, K., 98–100, 99f, 107t
Bebrone, C., 50–51, 53, 54
Beck, J. L., 61–62, 65–66
Becke, A. D., 98–100
Beintema, J. J., 3–5
Bellia, F., 118t, 121, 132–135, 134f
Bellott, M., 105–106
Bellotti, V., 116–120
Belviso, B. D., 124
Ben-Bassat, A., 57–58
Bender, K. M., 54
Benini, S. F. M., 70
Benitez, J., 71
Benkovic, S. J., 51–52
Bennett, B., 51–52, 53, 56–57, 58, 59–60
Bensman, T. J., 118t
Ben-Yosef, V. S., 5–6, 28, 28f
Bergantino, E., 6
Beringer, M. P., 118t
Berkecz, R., 121
Berkowitz, M. L., 84ge, 93
Berndt, M. C., 132–135
Bertini, I., 40, 40f, 41, 43–45, 44f, 46–47, 46f, 121, 123, 124–125
Besler, B. H., 96–97
Biancone, G., 7–8
Bianculli, A., 118t
Bida, L., 118t
Biertumpfel, C., 107t
Bihan, D., 41–44, 42f, 46, 47
Bijelic, A., 29
Bill, P., 65–66
Billeter, S. R., 86–87
Bisaglia, M., 6
Bischoff, R., 132–135
Bitler, A., 45–46
Blacher, S., 116–120
Blaha, G., 107t
Blanco-Labra, A., 5–6, 7, 23–27, 23f
Boccarelli, A., 124
Bode, W., 124–125
Boerner, F., 123–124
Bonaventura, C., 3–5

Bonaventura, J., 3–5
Bonnet, S., 116–120, 118t
Bonvoisin, J. J., 66–67
Boos, W., 60–61
Boosman, A., 57–58
Boresch, S., 85
Bortoluzzi, A. J., 61–62, 65–66, 67–68
Boschi, L., 53
Boucher, J.-L., 70–71
Bouillenne, F., 53
Bounaga, S., 54
Boutchard, C. L., 65–67
Bozzo, G. G., 65–66
Bradshaw, R. A., 57–58
Brahma, A., 101–102
Branca, F., 6
Brand, D. D., 45–46
Braun, M., 124–125
Braun, R., 86–87
Breece, R. M., 54
Brick, P., 38
Brooks, B. R., 86–87, 89–90
Brooks, C. L., 86–87, 89–90
Brown, C. A., 3
Broyde, S., 100–102
Bru-Martínez, R., 16
Brunak, S., 9–10
Bruyere, F., 116–120
Bucher, M., 65–66
Buchert, J., 5, 7, 15
Büchler, K., 5–6
Buerger, K., 118t
Büldt-Karentzopoulos, K., 7, 19, 20t
Buntkowsky, G., 123–124
Burgner, J. W., 116–120
Burt, S. K., 87–88, 88f, 89, 91–92
Burton, K. S., 16
Busby, S. A., 41–43, 131–132
Bush, A. I., 125–126, 132–135
Bush, K., 53
Bushnell, D. A., 103–106, 104f
Butt, L. E., 40, 41–43, 42f, 125–126
Byun, K. S., 85–86, 93, 96–97, 103–105

C

Cabanes, J., 16
Calderone, V., 41, 43, 44f, 46–47, 123, 124–125
Caldwell, J. W., 86–87, 93

Calheiros, R., 70–71
Caliandro, R., 124
Cama, E., 70–71
Camins, A., 118t
Campbell, H. D., 65–66
Caparrós-Martín, J. A., 118t
Carafoli, F., 41–44, 42f, 46, 47
Carfi, A., 51–52, 54
Carloni, P., 54–57
Carney, J. P., 61–62
Carpenter, J. E., 22, 23f
Carr, P. D., 61–62
Carrington, L. E., 61–62, 65–68, 67f
Carroll, B. J., 65–66
Carvajal, N., 70–71
Carvalho, A. T. P., 103–105
Casado-Vela, J., 16
Case, D. A., 86–87, 93, 96–97, 98–100
Casellato, A., 68
Cassady, A. I., 65–66
Cassady, I., 65–66
Castro, V., 71
Cavanagh, J., 121
Cavigliasso, G., 63–65
Ceccarelli, E. A., 56–57
Cerofolini, L., 40, 43–44, 45f
Cerpa, J., 70–71
Chalmers, M. J., 41–43, 131–132
Chambers, T. J., 65–66
Chandrasekar, S., 53, 56–57
Chang, S., 57–58
Chang, S. Y., 57–58
Chang, Y. H., 57–58, 59–60
Chapelon, C. G. J., 7–8
Cheatham, T. E., 93, 96–97, 98–100
Cheeseman, J. A., 98–100
Chen, H., 7–9
Chen, J., 128–129
Chen, L.-L., 57–58
Chen, Y., 116–120
Chen, Z., 121–122
Cheng, W., 7–9
Cheng, X., 116–120
Cheng, Y., 100–101
Chirichigno, J. W., 118t
Chisari, M., 6
Chiu, C. H., 59–60
Choi, I. S., 118t
Chojnacki, M., 17

Chung, L., 41
Chung, L. W., 86
Ciaccio, C., 118t, 129–131, 130f
Cirera, J., 18
Ciuraszkiewicz, J., 65–66
Ciurli, S., 70
Clarke, A. R., 53, 54
Clifton, M. C., 70–71
Cocklin, S. L., 118t
Coetzer, C., 6
Cofre, J., 71
Cohen, S. M., 121
Cohen, S. R., 40, 45–46
Coles, B., 65–66
Colige, A. C., 118t
Collier, I. E., 37–38
Coluccia, M., 124
Comba, P., 63–67, 70
Constabel, C. P., 5, 7, 8–10
Cook, J. O., 57–58
Cooke, N. E., 71
Copik, A. J., 57–58, 59–60
Corchado, J., 107–109
Cornacchia, D., 124
Correia, A., 53
Corsini, D., 6
Cosenza, M., 46–47, 124–125
Cosper, N. J., 58
Costello, A. L., 53, 54, 56–57
Cox, R. S., 68
Cox, T. M., 65–66
Craig, L., 61–62
Cricco, J. A., 56–57
Crichton, R. R., 116–120
Crowder, M. W., 50–52, 53, 56–57, 70
Cuff, M. E., 5–6, 22, 23f
Cui, Q., 105–106
Cui, Y.-M., 57–58
Cuniasse, P., 39–40, 41, 46
Curry, V. A., 38

D

da Silva, A. J. R., 6
D'Agata, R., 125–126, 132–135
Dahal, B., 118t
Dai, L., 128–129
Dal Peraro, M., 54–57, 55f
Dale, G. E., 58–59, 118t
Damblon, C., 51–52

Danciger, M., 118t
Dancis, A., 118t
Dancs, A., 121
D'Andrea, L., 132–135
Daniel, D., 132–135
Dapprich, S., 85–86, 93, 96–97, 103–105
D'Arcy, A., 58–59
Darden, T. A., 84ge, 93, 96–97, 98–100
Daumann, L. J., 50–52, 60–62, 63–65
Dauter, Z., 3–5
Davenport, R. C., 85
David, S. S., 66–67
Davis, J. C., 65–66
Day, E. P., 66–67
de Borst, G. J., 118t
de Jersey, J., 61–62, 65–68, 69
de la Pena, A., 65–66
de Pauw, E., 54
de Seny, D., 51–52, 54
de Souza Pessôa, G., 131–132
de Vries, J.-P. P. M., 118t
Decker, H., 3–6, 17, 29–30
del Pozo, J. C., 65–66
del Valle, J., 118t
Delbrück, H., 53
Deng, J., 5–6, 28, 28f, 29–30
Derbyshire, D., 123–124
Desimio, M. G., 118t
Detry, B., 116–120
Devreese, B., 53
Dewil, M., 118t
Di Cera, E., 121–122
Dick, A., 61–62, 65–66, 67–68
Dideberg, O., 53, 54
Dillinger, R., 19, 20t
Dinakarpandian, D., 41
Ding, L., 86
Dionysius, D. A., 65–66
Dirks-Hofmeister, M. E., 16
Dismukes, C. G., 50
Dive, V., 39–40, 41, 46
Diven, C., 57–58, 60
Doan, P. E., 66–67
Dobson, C. M., 116–120
Docquier, J.-D., 53
Dohler, F., 118t
Douangamath, A., 58–59
Douwe de Boer, A., 9

Drakenberg, T., 123
D'Souza, V. M., 57–58
Duee, E., 51–52, 54
Duez, C., 51–52, 54
Duff, A. P., 66–67
Duke, R. E., 96–97
Dunbrack, R. L., 105–106
Duncan, G., 118t
Dunham, W. R., 66–67
Dunlap, C., 94–95
Dunn, D. M., 57–58
Dunn, E. L., 65–66
Duran-Vilaregut, J., 118t
Durmus, A., 65–66

E

Ebbelaar, C. E. M., 7–8
Edwards, D. R., 118t
Edwards, T. E., 70–71
Ehrmann, M., 60–61
Eicken, C., 5–6, 7, 19, 20t, 22–23, 23f, 29–30, 65–66
Eickman, N. C., 3, 18, 19–22, 20t
Eisen, A. Z., 45
Ek-Rylander, B., 65–66
Elliott, T. W., 61–62, 65–66, 67–68, 67f
Elson, E. L., 37–38
Ely, F., 50–51, 53–54, 60–65
Emanuelsson, O., 9–10
Emery, D. C., 54
Emig, F. A., 70–71
Endo, T., 9
Engeland, C. G., 116–120
Enriquez, S., 71
Erickson, J. W., 87–88, 88f, 89, 91–92
Erpicum, C., 116–120
Escribano, J., 16
Esposito, S., 123
Essmann, Perera, L., 84ge, 93
Estiu, G., 54–56
Evans, L. H., 17–18, 19–22, 20t
Evanseck, J. D., 105–106
Evdokimov, A. G., 57–58, 60

F

Fabiane, S. M., 54
Faccio, G., 5, 7, 15
Fairbrother, W. J., 121
Fairweather, N., 57–58, 60

Falker, C., 118*t*
Farndale, R. W., 41–44, 42*f*, 46, 47
Fast, W., 51–52
Felici, A., 53
Felsenfeld, G., 20*t*, 22
Fennen, J., 86–87
Fernandes, P. A., 103–105
Feuerstein, R., 71
Fialho, E., 6
Field, M. J., 85, 105–106
Fields, G. B., 37–48, 125–126, 127–128, 128*f*, 131–132
Fischer, H., 19, 20*t*
Fishman, A., 5–6, 17, 28, 28*f*
Flanagan, J. U., 65–66
Florian, J., 97–98
Flurkey, W. H., 5, 7, 9, 16
Folkman, J., 57–58
Foloppe, N., 105–106
Foo, J. L., 60–61
Forgiarini, L. F., 118*t*
Fragai, F., 40*f*, 41, 43–45, 46*f*
Fragai, M., 40, 41, 43–44, 44*f*, 45*f*, 46–47, 121–122, 123, 124–125, 132–135
Franceschini, N., 54
Franklin, S. L., 51–52
Fraústo Da Silva, J. J. R., 116–120
Freedman, T. B., 3, 18, 22
Frenkel, J., 40
Frère, J.-M., 51–52, 53, 54, 56–57
Fridman, R., 125–126
Frieden, C., 121–122
Frisch, M. J., 98–100
Fröhlich, R., 65–67
Frossard, E., 65–66
Fuentealba, P., 71
Fujieda, N., 5–6, 9, 17–18, 29
Fujisawa, K., 3, 17–18
Fujita, Y., 8
Fukuchi-Mizutani, M., 3–5, 4*f*, 6, 9–10
Fukui, Y., 3–5, 4*f*, 6, 9–10
Funhoff, E. G., 65–66
Fusetti, F., 5–6, 28–30, 28*f*

G

Gahan, L. R., 50–52, 53–54, 60–68, 67*f*, 69, 70
Gajda, T., 121
Gajendrareddy, P. K., 116–120
Gallagher, L. A., 70–71
Galleni, M., 50–52, 53, 54, 56–57
Gamblin, S. J., 54
Gandia-Herrero, F., 16
Gao, J., 7–9, 105–106, 107–109
Garau, G., 50–51, 53, 54
Garcia, D., 71
Garcia, J. R., 71
Garcia, R. A., 131–132
Garcia-Carmona, F., 16
Garcia-Diaz, M., 98–100, 99*f*, 107*t*
García-Sáez, I., 50–51
Garcia-Viloca, M., 107–109
Gardberg, A. S., 70–71
Garg, R. K., 118*t*
Gasparetti, C., 5–6, 7, 15, 19, 20*t*, 23*f*, 27, 29–30
Gaykema, W. P. J., 3–5
Ge, Y., 65–66
Geacintov, N. E., 100–102
Geraldes, C. F. G. C., 40, 43–44, 45*f*
Gerdemann, C., 7, 19
Gerner, C., 19–22, 20*t*
Gerritsen, Y. A. M., 7–8
Gervasi, D. C., 125–126
Gessner, C. R., 131–132
Ghanem, E., 60–61
Ghosh, M., 57–58
Gil, F. P., 70–71
Ginsbach, J. W., 18
Giorgetti, S., 116–120
Giuffrida, M. L., 116–120
Go, K. V., 131–132
Gohlke, H., 93, 98–100
Gökmen, V., 6
Goldberg, G., 37–38
Goldfeder, M., 17
González, J. M., 53, 56–57
Gonzalez, L. J., 54
Gordon, D. M., 118*t*
Gossas, T., 125–126, 127*f*
Gozzo, F. C., 131–132
Grasso, G., 115–142
Gregory, M., 107*t*
Griffin, P. R., 41–43, 131–132
Grinius, L., 57–58, 60
Groenhof, G., 85

Groothaert, M. H., 18
Grossman, M., 41
Grossmann, J. G., 40
Grunden, A. M., 57–58
Gruszczynska-Biegala, J., 131
Guddat, L. W., 50, 53–54, 60–63, 62f, 65–68, 69, 70
Gumbart, J., 86–87
Guo, L., 8
Guo, Z., 116–120
Gupta, V., 118t
Gwag, B. J., 118t

H

Ha, J. Y., 70–71
Hadler, K. S., 49–82, 62f
Haftek, M., 118t
Hajdin, C., 53, 56–57
Hajdin, C. E., 54
Hakulinen, N., 5–6, 15, 19, 20t, 23f, 27, 29–30
Halaouli, S., 6, 8
Halbwirth, H., 3–5
Hamdi, M., 6, 8
Hamilton, S., 61–62, 65–68
Hamilton, S. E., 65–67
Hampel, H., 118t
Han, M. V., 11–12, 12f
Han, S., 70–71
Hanaoka, K., 127–128
Hansen, D. F., 121–122
Hanson, G. R., 62–67, 68, 70
Hanzlik, R. P., 58–60
Haraguchi, K., 65–66
Harel, E., 3–5
Hartmann, H., 17
Hase, A., 65–66
Hashmi, N., 129
Hass, M. A. S., 121–122
Hasty, K. A., 37–38
Hatana, S., 8
Hatano, M., 121–122
Haussman, G., 65–66
Hawk, M. J., 54
Hay, P. J., 98–100
Hay, S., 107–109
Hayashi, H., 6
Hayman, A. R., 65–66
Hazes, B., 3–5

He, H., 121
Heine, A., 58–59
Heinz, U., 50–52, 53, 54, 56–57
Hemmingsen, L., 51–52
Hendrickson, W. A., 5–6, 22, 23f
Hendriks, M., 7–8
Hengge, A. C., 68
Henkin, J., 57–58
Henriques, I., 53
Henriquez, A., 71
Heppner, D. E., 18
Hermans, J., 132–135
Hernandez-Valladares, M., 51–52, 53
Herrera, P., 70–71
Himmelwright, R. S., 3, 18, 19–22, 20t
Hirao, H., 86
Hodgkinson, L. M., 118t
Hoeberichts, F. A., 7–8
Hoerning, K. R., 57–58
Hoffman, B. M., 66–67
Hol, W. G. J., 3–5, 22
Holz, R. C., 57–58, 59–60
Honek, J. F., 57–60
Hooper, N. M., 116–120
Höpfner, K. P., 61–62
Horan, M. P., 116–120
Hou, C., 50–51
Howatt, D. A., 118t
Hu, H., 102
Hu, L., 87
Hu, W., 122
Hu, X., 58–59
Huang, M., 57–60
Huang, Q.-Q., 57–58
Huang, W. J., 57–60
Huang, X., 103–105
Huang, Y., 116–120
Huber, R., 124–125
Hudson, P. S., 85
Humbert, P., 118t
Hume, D. A., 61–62, 65–68
Hummer, G., 86–87
Hunenberger, P. H., 86–87
Hupe, D. J., 122

I

Ichishima, E., 16–17
Ichisima, E., 8

Ijaz, T., 118*t*
Ikeda, T., 5–6, 9, 17–18, 29
Ilag, L. L., 132–135
Ilangovan, U., 121
Ilchenko, S., 118*t*
Inlow, J. K., 5, 7, 9, 16
Iqbal, A., 129
Isaksson, J., 123–124
Ismaya, W. T., 5–6, 28–30, 28*f*
Iyer, S., 38–39, 39*f*

J

Jackson, C. J., 60–63, 62*f*
Jaenicke, E., 5–6, 29–30
Jaiswal, R., 123
Jancso, A., 121
Jarvet, J., 116–120
Jarvi, M., 128–129
Jeffrey, J. J., 45
Jenkins, N., 118*t*
Jensen, M., 51–52
Jensen, R. M., 121–122
Jensen-Taubman, S., 116–120
Jiang, H., 5–6, 28, 28*f*, 29–30
Jimenez-Atienzar, M., 16
Joachimiak, A., 116–120
Johansson, E., 61–62, 66–67
Johnson, J. L., 118*t*
Johnson, L. L., 122
Johnson, R. E., 101–102
Johnston, E. M., 18
Jolley, R. L. Jr., 17–18, 19–22, 20*t*
Junges, R., 116–120

K

Kabeya, M., 125–126
Kaija, H., 61–62, 66–67
Kaintz, C., 1–36
Kakuta, D., 8
Kaljunen, H., 5–6, 15, 19, 20*t*, 23*f*, 27, 29–30
Kalk, K. H., 3–5
Kamerin, S. C. L., 97–98
Kampatsikas, I., 3–5
Kang, F., 116–120
Kanteev, M., 5–6, 17, 28, 28*f*
Kaplan, C. D., 103–106, 104*f*
Karcher, A., 61–62

Karplus, M., 85
Kawai, M., 57–58
Kawamura, K., 9
Kawamura-Konishi, Y., 8
Kawano, T., 8
Keough, D. T., 65–66
Kerff, F., 53
Ketcham, C. M., 65–66
Khalil, R. A., 116–120
Khattri, S., 118*t*
Kiefer, M., 51–52, 54
Kiesewetter, D. O., 127–128
Kikuchi, S., 3–5, 4*f*, 6, 9–10
Kim, D., 70–71
Kim, J., 132–135
Kim, K. H., 70–71
Kiselar, J., 118*t*
Kitajima, N., 3, 17–18
Klaassen, C. H., 65–66
Klabunde, T., 5–6, 22–23, 23*f*, 29–30, 61–62, 65–67
Klanke, C., 57–58, 60
Klebe, G., 58–59
Klein, C. D. P., 58–59
Klein, M. L., 54–57
Klingler, L. J., 57–58
Klinkenberg, M., 59–60
Klinman, J. P., 107–109
Klippenstein, S. J., 107–109
Knispel, R., 118*t*
Knöfel, T., 61–62
Knowles, V. L., 65–66
Kocourek, A., 124–125
Kohen, A., 107–109
Kollman, P. A., 86–87, 93, 96–97
Komaromi, I., 85–86, 93, 96–97, 103–105
Komatsu, T., 127–128
Kondo, Y., 107*t*
Konig, G., 85
Konigsberg, W. H., 107*t*
Kornberg, R. D., 103–106, 104*f*
Korsinczky, M., 65–66
Kortz, U., 5–6, 28–29
Kosaka, K., 118*t*
Kossmann, J., 65–66
Kovacs, H., 123–124
Kowalski, J., 51–52
Kragl, M., 9

Krahn, J. M., 87, 88f, 89–94, 91f, 92f, 98–101, 99f, 107t
Krasemann, S., 118t
Kraut, J., 87, 88f, 89–90
Kraynov, V. S., 89–90, 94–95
Krebs, B., 5–6, 7, 19, 20t, 22–23, 23f, 29–30, 61–62, 65–67
Kruus, K., 5–6, 7, 8, 15, 19, 20t, 23f, 27, 29–30
Kuchta, R. D., 97–98
Kumagai, T., 5–6, 27, 29–30
Kunkel, T. A., 98–100, 99f, 107t
Kupper, M. B., 53
Kupper, U., 8
Kurisu, G., 5–6, 9, 29
Kurochkin, A. V., 122
Kurokawa, M., 118t
Kuryavyi, V., 100–101
Kuznetsov, A. M., 107–109

L

La Mendola, D., 121
Lamarche, B., 94–95
Lambert, C. A., 118t
Lamotte-Brasseur, J., 53
Lang, G., 118t
Lang, L., 127–128
Lang, R., 124–125
Langley, D. B., 66–67
Laraki, N., 54
Larrabee, J. A., 19, 57–58, 61–65
Larson, T. J., 60–61
Lau, T. C., 67–68
Lauer, J. L., 40f, 41, 43–45, 46f, 121
Lauer-Fields, J. L., 41–43, 131–132
Laws, A. P., 54
Lebrun, I., 132–135
Led, J. J., 121–122
Lee, B. R., 16–17
Lee, C., 98–100
Lee, C. H., 132–135
Lee, C. Z., 59–60
Lee, H., 84ge, 93, 127–128, 128f
Lee, H. H., 70–71
Lee, H. J., 132–135
Lee, J., 70–71
Lee, J. K., 118t
Lee, J. Y., 107t

Lee, S., 94–95, 127–128
Lee, S. P., 118t
Lee, T., 86–87, 98–100, 101–102
Lee, Y.-M., 46–47, 124–125
Lentz, D., 123–124
Leopoldini, M., 57–58
Lerch, K., 3, 8, 16, 18, 19–22, 20t
Leung, E., 61–62, 65–66, 67–68, 67f
Leung, E. W. W., 50–51, 53–54, 65–66
Levitt, M., 85
Leyva, A., 65–66
Li, D., 116–120
Li, H., 128–129
Li, J., 38, 57–58, 70–71
Li, J.-Y., 57–58
Li, X., 86
Li, Y., 5–6, 28, 28f, 29–30, 60–61, 116–120
Liang, J., 118t
Liebhaber, S. A., 71
Lim, S., 57–59, 59f
Lin, G.-Q., 125–126, 127f
Lin, H., 85
Lin, K. S., 59–60
Lin, L. Y., 59–60
Lin, P., 87, 89–97, 92f, 98–100, 102–103, 107–109, 107t
Lin, S. S., 65–66
Lindqvist, Y., 61–62, 66–67
Ling, C., 59–60
Linzen, B., 3–5
Lior-Hoffmann, L., 101–102
Lipscomb, W. N., 61–62
Lisa, M. N., 53, 56–57
Liu, E., 132–135
Liu, J., 116–120
Liu, J. O., 58–59
Liu, J. W., 61–62
Liu, P., 70–71
Liu, X., 7–9, 103–105
Ljusberg, J., 65–66
Llarrull, L. I., 51–52, 54–56, 55f
Lloyd, L. F., 38
Loaiza, A., 116–120
Lobos, M., 71
Loehr, J. S., 3, 18, 22
Loehr, T. M., 3, 18, 22
Lomascolo, A., 6
Lone, S., 101–102

Lonhienne, T., 61–62
Lopes, M. L. M., 6
López, V., 70–71
Lorenzl, S., 118*t*
Lou, P., 57–58
Love, S., 6
Lowery, M. D., 2–3, 17–18
Lowther, W. T., 57–60, 59*f*
Lu, J. P., 57–58, 59–60
LuBien, C. D., 3, 18, 19–22, 20*t*
Luchinat, C., 40, 40*f*, 41, 43–45, 44*f*, 45*f*, 46–47, 46*f*, 121–122, 123, 124–125
Luo, J., 116–120
Luo, Q.-L., 57–58
Luo, R., 93, 98–100

M

Ma, X., 7–9
Ma, Y., 127–128
Ma, Z.-Q., 57–60
Macholl, S., 123–124
Mackerell, A. D., 86–87, 89–90, 105–106
MacSweeney, A., 58–59
Madden, D. T., 57–59, 59*f*
Magnus, K. A., 3–5, 22, 23*f*
Magrì, A., 121
Magrini, A., 7, 19
Maillard, C., 116–120
Makino, N., 19–22, 20*t*
Maletta, M., 43, 44*f*, 124–125
Malgieri, G., 118*t*, 121–122, 123
Malinina, L., 100–101
Mammi, S., 6
Manganello, J. M., 118*t*
Manich, G., 118*t*
Manka, S. W., 41–44, 42*f*, 46, 47
Marasco, D., 129–131, 130*f*
Mark, A. E., 86–87
Markl, J., 5–6
Marmer, B. L., 37–38
Marques, M. P., 70–71
Marshall, K., 65–66
Marshall, M. R., 6
Martin, J. L., 61–62, 66–68
Martin, L. B., 17
Martinelli, N. C., 118*t*
Martinez, M. V., 6
Martínez-Glez, V., 118*t*

Martinez-Oyanedel, J., 71
Marucha, P. T., 116–120
Marusek, C. M., 5, 7
Mason, H. S., 17–18, 19–22, 20*t*
Mathews, S., 121–122
Matoba, Y., 5–6, 27, 29–30
Matrisian, L. M., 127–128
Matthews, B. W., 57–60, 59*f*
Mauch, C., 118*t*
Mauracher, S. G., 1–36
Mayer, A. M., 3–5, 10
Mayer, R. L., 19–22, 20*t*
McAlpine, A. S., 61–62, 66–68
McCarthy, B. Y., 61–62
McConachie, L. A., 65–66
McCormick, J. M., 66–67
McGeary, R. P., 50–51, 53–54, 60–61, 65–67, 69, 70
McIntyre, J. O., 127–128
McKibbin, M., 118*t*
Medina, S., 118*t*
Meisinger, T., 118*t*
Meiwes, D., 19, 20*t*
Melikian, M., 40, 40*f*, 41, 43–45, 46*f*, 121, 123
Mendiola-Olaya, E., 5–6, 7, 23–27, 23*f*
Meng, L., 57–58
Mercuri, P. S., 50–51, 53
Merkx, M., 65–66, 68
Merz, K. M., 54–56, 93, 96–97, 98–100
Mes, J. J., 5–6, 7–8, 28–30, 28*f*
Meunier, C., 54
Meurer, L., 118*t*
Meyer, H. E., 7, 19
Meyer-Klaucke, W., 51–52, 54
Meyers, R. A, 50
Miao, D., 132–135
Michael, C., 3–5, 9, 19–22, 20*t*
Mikhailova, M., 121
Miller, D., 70–71
Miller, J. H., 89, 91–92
Miller, K. I., 5–6, 22, 23*f*
Minond, D., 41–43, 131–132
Miraula, M., 49–82, 51*f*
Mitchell, T., 54
Mitić, N., 49–82, 51*f*, 62*f*, 67*f*
Mitternacht, S., 41
Mobashery, S., 125–126

Moerschbacher, B. M., 16
Molderings, G. J., 70–71
Molitor, C., 3–6, 9, 19–22, 20t, 28–29
Molloy, S., 17
Montes, P., 71
Montgomery, J. A., 85–86, 93, 96–97, 103–105
Moore, B. M., 16
Moore, M. D., 3–5
Moorleghen, J. J., 118t
Morán-Barrio, J., 53, 56–57
Moreno, A., 5–6, 7, 23–27, 23f
Morita, N., 65–66
Moriyama, H., 132–135
Moro, N., 118t
Morokuma, K., 85–86, 93, 96–97, 103–105
Moro-oka, Y., 3, 17–18
Moss, L. A. S., 116–120
Moubaraki, B., 65–67
Mu, Y., 116–120
Mulholland, A. J., 87, 97–98
Muraki, N., 5–6, 9, 29
Murata, M., 5, 17–18
Murray, K. S., 65–67
Murray, T. P., 61–62
Musharaf, S. G., 129
Myambo, K., 57–58
Mylonas, E., 40
Myochin, T., 127–128

N

Nagano, T., 127–128
Nagase, H., 38–40, 39f, 41, 46
Nagy, N. V., 121
Nair, D. T., 101–102
Nakai, M., 9
Nakajima, T., 16–17
Nakamura, M., 16–17
Nakamura, Y., 5, 17–18
Nakayama, M., 132–135
Nakayama, T., 3–5, 4f, 6, 9–10
Nakazato, H., 65–66
Nam, K., 105–106
Nan, F.-J., 57–58
Narvaes, L. B., 118t
Nash, K., 65–66
Natile, G., 124
Neira, B., 71

Neves, A., 66–67
Nguyen, A. N., 118t
Nicholson, R. I., 116–120
Nick, H. S., 65–66
Niedermann, D. M., 8
Nielsen, H., 9–10
Nikodinovic-Runic, J., 17
Nillius, D., 29–30
Nilsson, I., 86–87, 89–90
Nilsson, S., 65–66
Nishikoori, M., 65–66
Niu, W., 121–122
Nkyimbeng, T., 118t
Noble, C. J., 65–67, 68
Nordstroem, H., 125–126, 127f
Norgard, M., 65–66
Nusgens, B. V., 118t
Nushi, F., 124
Nuttleman, P. R., 65–66
Nuvolone, M., 116–120
Nystroem, S., 123–124

O

Oakley, A. J., 61–62
Obici, L., 116–120
Oda, M., 118t
Oddie, G. W., 69
Oefner, C., 58–59, 118t
O'Halloran, T. V., 51–52
O'Hare, M. C., 38
Ohba, Y., 16–17
Okamoto, T., 65–66
Okuyama, H., 65–66
Olczak, M., 65–66
Olczak, T., 65–66
Ollis, D. L., 50–52, 51f, 53–54, 56–57, 60–67, 70
Olson, A. C., 97–98
Olson, M. W., 125–126
Opdenakker, G., 45–46, 118t
Orellana, M. S., 71
Orellano, E. G., 56–57
Orlandi, A., 118t
Orringer, E. P., 118t
Orville, A. M., 57–59, 59f
Oshima, M., 118t
Outten, C. E., 51–52
Oyama, T., 5–6, 9, 29

P

Paal, K., 132–135
Page, M. I., 50–51, 54, 56–57, 70
Pain, D., 118t
Pal, S., 121
Palladini, G., 116–120
Pallas, M., 118t
Palmer, A. G., III., 121
Palumaa, P., 116–120
Pande, V. S., 107–109
Pant, K. K., 118t
Pantazatos, D. P., 131–132
Pappalardo, G., 129–131
Pares, S., 51–52, 54
Parigi, G., 40, 43–44, 45f, 121–122
Park, C., 57–58
Park, S. M., 132–135
Parkin, E. T., 118t
Paronen, J., 132–135
Parr, R. G., 98–100
Parry, D. A., 118t
Partis, M. D., 16
Pate, J. E., 18
Patrapuvich, R., 70–71
Patro, J. N., 97–98
Pauland, L. N., 116–120
Paul-Soto, R., 51–52, 56–57
Paupert, J., 116–120
Pavek, J., 6
Payne, D. J., 54
Pedersen, L. C., 87, 88f, 89–97, 91f, 92f, 98–101, 99f, 102–103, 107–109, 107t
Pedersen, L. G., 83–114, 84ge
Pedone, C., 132–135
Pedroso, M. M., 49–82
Peixe, L., 53
Pelletier, H., 87, 88f, 89–90
Peng, S., 128–129
Peng, W.-J., 116–120
Peralta, R. A., 61–62, 65–66, 67–68
Perera, L., 83–114
Pervushin, K., 121
Peterson, J., 66–67
Pethe, S., 70–71
Petrella, R. J., 86–87, 89–90
Petsko, G. A., 85
Pettit, M., 53
Phelan, E., 50–51, 53–54
Phelan, E. K., 50–52, 51f, 56–57, 70

Phillips, D. R., 118t
Phillips, J. C., 86–87
Pi, Y., 128–129
Pickford, A. R., 40, 41–43, 42f, 125–126
Pierau, S., 58–59, 118t
Pietropaolo, A., 121, 129–131, 132–135, 134f
Pilau, E. J., 131–132
Piletz, J. E., 70–71
Pimenta, D. C., 132–135
Pinkse, M. W., 65–66
Plaxton, W. C., 65–66
Ponder, J., 98–100
Prasad, B. R., 97–98
Prasad, R., 87, 88f, 89–90
Prely, L. M., 132–135
Pringle, J. A., 65–66
Printz, M. P., 20t, 22
Prior, S. H., 40, 41–43, 42f, 125–126
Puerta, D. T., 121
Puig, B., 118t
Pulina, M. O., 118t

Q

Qayyum, M., 18
Qu, B.-H., 118t
Quaroni, L., 66–67
Que, L., Jr., 66–67
Queiroz, C., 6

R

Radhakrishnan, R., 89–90, 100–101
Raetz, C. R., 60–61
Raffetto, J. D., 116–120
Raghothama, K. G., 65–66
Rajic, A., 132–135
Ramon-Maiques, S., 107t
Ramos, M. J., 103–105
Randall, C. R., 66–67
Rao, A. P., 118t
Rasia, R. M., 56–57
Rastogi, V., 57–58, 60
Raushel, F. M., 60–61
Ravera, E., 121–122
Raynal, N., 41–44, 42f, 46, 47
Read, C. M., 40, 41–43, 42f, 125–126
Rechkoblit, O., 100–101
Recourt, K., 7–8
Regierer, B., 65–66

Reilly, B., 123
Renda, M., 123
Reyes Grajeda, J. P., 5–6, 7, 23–27, 23f
Reymond, J. L., 127–128
Ribbe, A., 116–120
Rich, D. H., 57–59, 59f
Riek, R., 121
Riggs, A. F., 3–5
Riley, M. J., 68
Ringe, D., 85
Rittenhouse, R. C., 89, 91–92
Rival, S., 51–52, 53
Rizzarelli, E., 116–120, 121, 125–126, 129–131, 132–135
Rizzi, A., 9
Robb, M. A., 98–100
Robberecht, W., 118t
Roberts, R. M., 65–66
Robertson, J. V., 118t
Robichaud, T. K., 41
Robinson, N. J., 9
Rod, T. H., 87
Rodgers, K. R., 121
Rodrigues, A. L. S., 70–71
Roitberg, A. E., 97–98
Rojas, I. G., 116–120
Rompel, A., 1–36
Ronau, J. A., 116–120
Root, D. E., 3, 18
Rosenberg, G. A., 118t
Rosenblum, G., 40, 45–46
Rosner, M. R., 116–120
Rossolini, G. M., 53, 54
Rosta, E., 86–87
Rothlisberger, U., 54–56, 55f
Rougier, A., 118t
Rouvinen, J., 5–6, 15, 19, 20t, 23f, 27, 29–30
Roux, B., 86–87, 89–90
Rozeboom, H. J., 5–6, 28–30, 28f
Rubio, V., 65–66
Ruebush, S., 57–58
Ruppert, C., 118t
Russo, L., 123
Russo, N., 57–58
Russo, P., 116–120
Ruvo, M., 129–131, 130f
Ryde, U., 87
Ryu, K. S., 118t

S

Saavedra, M. J., 53
Sacchettini, J. C., 5–6, 22–23, 23f, 29–30
Saffarian, S., 37–38
Saito, M., 118t
Sala-Newby, G. B., 118t
Salas, M., 70–71
Salluzzo, A., 123
Saloheimo, M., 5, 7, 15
Salomone, F., 129–131
Samorì, B., 6
Sampson, P. B., 57–60
Samyn, B., 65–66
Sanchez, M. L., 107–109
Sands, R. H., 66–67
Santucci, P., 54
Saraiva, M. J., 118t
Sarti, N., 40
Sasanelli, R., 124
Sato, T., 3–5, 4f, 6, 9–10
Satriano, J., 70–71
Sawaya, M. R., 87, 88f, 89–90
Sbardella, D., 118t
Schartau, W., 3–5
Schenk, G., 49–82, 51f, 67f
Scherer, R. L., 127–128
Schiffmann, R., 58–59
Schlegel, H. B., 98–100
Schlick, T., 89–90, 97–98, 100–101
Schliemann, W., 6
Schlömer, P., 53
Schneider, G., 61–62, 66–67
Schneider, H. J., 3–5
Scholte, M., 66–67
Scholtes, V. P. W., 118t
Schoonheydt, R. A., 18
Schoot Uiterkamp, A. J., 19
Schröder, G. F., 5–6
Schulz, H., 58–59, 118t
Schurmans, C., 118t
Scott, W. R. P., 86–87
Scrutton, N. S., 107–109
Scuseria, G. E., 98–100
Searle, I. R., 65–66
Seeger, W., 118t
Selleck, C., 49–82, 51f
Sellés-Marchart, S., 16
Sels, B. F., 18
Sendovski, M., 5–6, 28, 28f

Senn, H. M., 85
Sertchook, R., 40
Sharma, N. P., 53, 56–57
Shen, Y., 116–120
Sheppard, G. S., 57–58
Shi, Q., 118*t*
Shimizu, T., 121–122
Shimokawa, C., 5, 17–18
Shin, J. H., 118*t*
Shiomi, T., 118*t*
Shock, D. A., 94–95
Shock, D. D., 87, 88*f*, 89–94, 91*f*, 92*f*, 95–97, 98–101, 102–103, 107–109, 107*t*
Shoji, A., 125–126
Shu, Q., 121–122
Sigoillot, J.-C., 6, 8
Sikora, K., 132–135
Siliqi, D., 124
Silva, D. A., 103–105
Silvello, D., 118*t*
Simmen, R. C., 65–66
Simmerling, C. L., 96–97
Singh, M. K., 118*t*
Siwakoti, A., 118*t*
Skarzynski, T., 38
Skelton, N. J., 121
Smeets, P. J., 18
Smith, A. I., 132–135
Smith, G. M., 5–6, 7, 23–27, 23*f*
Smith, G. N., Jr., 37–38
Smith, J. A., 57–58
Smith, S. J., 66–67, 68
Smoukov, S. K., 66–67
Snedden, W. A., 65–66
Sobol, R. W., 85
Soderhjelm, P., 87
Soederberg, K. L., 116–120
Soeter, N. M., 3–5
Sohi, M. K., 54
Sokolov, A. V., 118*t*
Solano, F., 17
Solomon, E. I., 2–3, 17–18, 19–22, 20*t*, 66–67
Song, D., 71
Sorensen, D., 118*t*
Sousa, J. C., 53
Spagna, G., 6
Spencer, J., 50–51, 53, 54, 70
Spener, F., 7, 19, 65–66

Spiro, T. G., 19
Spoto, G., 116–120, 125–126, 128–131, 130*f*, 132–135
Stanciu, L., 116–120
Starckx, S., 118*t*
Steffensen, B., 41
Stefflova, K., 128–129
Stetler-Stevenson, W. G., 116–120
Stewart, J. J. P., 103–105
Stoczko, M., 53
Straatsma, T. P., 89, 91–92
Strack, D., 6
Strange, R. W., 65–66
Stranger, R., 63–65
Sträter, N., 61–62, 65–67
Strickland, E., 118*t*
Stura, E. A., 39–40, 41, 46
Suarez, D., 54–56
Subramanian, V., 118*t*
Sudlow, G. P., 127–128, 128*f*
Sugawara, M., 125–126
Sugiyama, M., 5–6, 27, 29–30
Suh, J. H., 118*t*
Summors, A. C., 65–66
Sun, H. B., 37–38
Sun, Z.-H., 125–126, 127*f*
Sung, J. Y., 132–135
Swierczek, S. I., 58, 59–60
Sykes, A. G., 67–68
Szpoganicz, B., 61–62, 65–66, 67–68

T

Tainer, J. A., 61–62
Tajhorshid, E., 86–87
Tam, M. F., 59–60
Tanaka, Y., 6
Tang, W.-J., 116–120
Tanifum, E. A., 61–63, 62*f*
Tao, J.-H., 116–120
Taylor, D. R., 118*t*
Taylor, I. A., 54
Taylor, J. S., 5, 7, 8–10
Taylor, K. M., 116–120
Tcherkalina, O. S., 118*t*
Teichert, U., 57–58
Tempera, G., 70–71
Temtamy, S., 118*t*
Tenorio, J., 118*t*
Terai, T., 127–128

Tessari, I., 6
Thiel, W., 85
Thill, J., 3–5
Thomas, P. J., 118t
Thormann, M., 58–59
Thurm, D. K., 118t
Thurston, C. F., 16
Tiiman, A., 116–120
Tioni, M. F., 51–52
Tironi, I. G., 86–87
Titman, J. J., 123–124
Toccafondi, M., 40f, 41, 43–45, 46f, 121
Ton-That, H., 22, 23f
Toomes, C., 118t
Toriumi, K., 3, 17–18
Torres, C., 70–71
Toscano, M., 57–58
Tougu, V., 116–120
Towns, K. V., 118t
Townson, S. A., 101–102
Tran, L. T., 5, 7, 8–10
Travaglini, G., 8
Treml, A., 58–59
Trobaugh, N. M., 5, 7
Trucks, G. W., 98–100
True, A. E., 66–67
Truhlar, D. G., 85, 107–109
Tsai, M.-D., 89–90, 94–95, 97–98
Tschesche, H., 124–125
Tsuji, M., 8
Tsunasawa, S., 57–58
Tsutsui, K., 118t
Tucker, P., 66–67
Tuczek, F., 3–5, 19, 20t
Tumer, N., 6
Tundo, G. R., 118t, 129–131, 130f
Tunon, I., 97–98
Turner, A. J., 118t
Twitchett, M. B., 67–68

U

Uchida, H. A., 118t
Udi, Y., 41
Uljon, A. N., 101–102
Ullah, J. H., 54
Ulstrup, J., 107–109
Um, J. W., 132–135
Uraga, Y., 8

Urban, M., 97–98
Uribe, E., 49–82

V

Vadas, M., 132–135
Valencia, M., 118t
Valente-Mesquita, V. L., 6
Valizadeh, M., 65–67
Valle, F., 6
Vallee, B. L., 116–120
Vallejos, A., 71
Van Beeumen, J., 65–66
Van Den Bosch, L., 118t
Van den Steen, P. E., 40, 45–46
van der Heide, S., 132–135
Van der Kamp, M. W., 87
Van Doren, S. R., 122
van Gelder, C. W. G., 16
van Holde, K. E., 5–6, 22, 23f
van Oosterhout, A. J. M., 132–135
Vance, M. A., 18
Vanderbeld, B., 65–66
Vasilyev, V. B., 118t
Veljanovski, V., 65–66
Vella, P., 50–51, 53–54, 69
Vereijken, J. M., 3–5
Verma, C. S., 54
Viale, A. M., 53, 56–57
Viceconte, N., 70–71
Vihko, P., 61–62, 66–67
Vila, A. J., 50–52, 53, 54–57, 70
Vilaplana, J., 118t
Villa, E., 86–87
Villarreal, F. J., 131–132
Virador, V. M., 5–6, 7, 23–27, 23f
Visse, R., 38–40, 39f, 41–44, 42f, 46, 47
Volbeda, A., 22
von Heijne, G., 9–10
Vreven, T., 85–86, 93, 96–97, 103–105

W

Wadman, S., 58–59
Wadt, W. R., 98–100
Waermlaender, S., 116–120
Walker, K. W., 57–58
Wallberg, H., 123–124, 125–126, 127f
Walsh, N., 69
Walsh, T. R., 53, 54

Wan, T., 54
Wan,, Y.-N., 116–120
Wang, B.-X., 116–120
Wang, D., 103–106, 104f
Wang, J., 57–58, 86–87, 93, 96–97, 107t
Wang, L., 101–102, 118t
Wang, M., 107t
Wang, S., 101–102
Wang, W., 86–87, 118t
Wang, X., 66–67, 71
Wang, Y., 5–6, 28, 28f, 29–30, 94–95, 100–101, 127–128
Wang, Z., 51–52, 116–120
Warburton, M. J., 65–66
Warshel, A., 85, 97–98
Washio, K., 65–66
Waterson, S., 59–60
Watt, S. J., 61–62
Weber, G., 53
Wegmann, D. R., 132–135
Weijn, A., 5–6, 7–8, 28–30, 28f
Weisbeek, P. J., 9
Weiss, R., 57–58
Welgus, H. G., 45
Werneburg, B. G., 89–90, 94–95
West-Mays, J. A., 118t
Westover, K. D., 103–106, 104f
Whitaker, J. R., 6
White, R. H., 70–71
Wichers, H. J., 5–6, 7–8, 16, 28–30, 28f
Wider, G., 121
Williams, R. J. P., 116–120
Willmann, C., 53
Wilson, B. C., 128–129
Wilson, B. E., 65–66
Wilson, S. H., 83–114
Winge, D. R., 9
Witzel, H., 65–67
Woertink, J. S., 18
Wolf, R. M., 86–87, 93
Wommer, S., 51–52, 53, 54
Woo, T. T., 61–62
Wood, D., 131
Wood, D. A., 16
Woodcock, H. L., 85
Woods, V. L. Jr., 131–132
Wormstone, I. M., 118t
Worth, J. M., 118t

Wu, J., 7–9
Wu, S., 85, 97–98
Wüthrich, K., 121
Wynne, C. J., 65–66

X

Xia, S., 107t
Xia, X., 116–120
Xie, S.-X., 57–60
Xiong, W., 118t
Xu, B., 127–128, 128f
Xu, C., 60–61
Xu, H., 70–71
Xu, M.-H., 125–126, 127f
Xu, W., 116–120
Xu, X., 121

Y

Yabuta, S., 5–6, 9, 17–18, 29
Yamamoto, A., 5–6, 27, 29–30
Yamauchi, S., 16–17
Yan, J.-W., 116–120
Yang, G.-J., 116–120
Yang, K.-W., 53, 56–57
Yang, W., 86–87, 98–100, 101–102
Yang, Y.-S., 66–67
Ye, Q.-Z., 57–60, 122
Yeo, K. J., 43, 44f, 124–125, 132–135
Yi, E., 57–58
Yi, H., 131
Yip, S. H., 61–65, 62f
Yiu, D. T., 67–68
Yokota, H., 37–38
Yonekura-Sakakibara, K., 3–5, 4f, 6, 9–10
York, D. M., 84ge, 93, 105–106
Yoruk, R., 6
Yoshida, N., 41
Yoshitsu, H., 5–6, 27, 29–30
Yu, A., 121

Z

Zakharova, E. T., 118t
Zayed, M., 118t
Zekiri, F., 19–22, 20t, 29
Zeppezauer, M., 51–52, 53, 56–57
Zerner, B., 65–66
Zezzi Arruda, M. A., 131–132
Zhang, J., 116–120

Zhang, J. Z. H., 116–120
Zhang, R., 105–106, 106f
Zhang, X., 128–129
Zhang, Y., 57–60, 86–87, 98–100, 101–102
Zhao, Y., 107t
Zhao, Z., 131
Zheng, G., 128–129
Zheng, Q., 128–129
Zhenxin, H., 54
Zhong, X., 89–90, 94–95
Zhou, Y., 116–120
Zhu, J., 128–129
Zhu, L., 127–128
Zigrino, P., 118t
Zimmermann, P., 65–66
Zippel, F., 7, 19, 20t
Zmasek, C. M., 11–12, 12f
Zolkiewska, A., 131
Zovinka, E. P., 51–52

SUBJECT INDEX

Note: Page numbers followed by "*f*" indicate figures and "*t*" indicate tables.

A

Aeromonas hydrophila, MβLs, 51*f*
Agaricus bisporus
  PPO3 of, 28–29
  tyrosinase sequences of, 15
Aggregation-prone proteins, 116–120, 118*t*
Agmatinase, 69–71
AmAS1. *See* Aureusidin synthase (AmAS1)
Amyloid fibrils, 116–120
Antibiotic resistance, MβLs, 50–57
Anticancer drugs, MetAP, 57–60
AoCO4. *See* Catechol oxidase from *Aspergillus oryzae* (AoCO4)
Arg254, 95–96
Asp190, 88
Asp256
  and catalytic magnesium ion, 89, 91–92
  hydrogen-bonded to, 89–90
  insertion reaction, 87
  Lewis base at, 107–109
  O3' proton to, 91–92
  oxygen atom, 91–92, 95–96
  in Pol β, 98–100, 101–103
Atomic force microscopy (AFM) analysis, 40
AuNPs. *See* Gold nanoparticles (AuNPs)
Aureusidin synthase (AmAS1), 3–5
  oxo complex, 22
  reaction catalyzed by, 4*f*
  in snapdragon petals, 5
  in vacuole, 9–10

B

Bacillus cereus
  BcII from, 54
  cefotaxime hydrolysis by, 55*f*
  MβLs, 51–52, 51*f*
Bacterial PPOs, 17
Bacterial tyrosinase, 27
BcII
  from *B. cereus*, 54
  cefotaxime hydrolysis by, 55*f*
  MβLs, 51–52, 51*f*

Binuclear metallohydrolases
  mechanistic diversity, 70–71
  reaction mechanism, 54–56, 56*f*, 59*f*
Bioremediator, glycerophosphodiesterase, 60–65
Blue copper proteins, 2–3
BmTYR. *See* Tyrosinase from *Bacillus megaterium* (BmTYR)

C

Caddie protein, 27
Catalytic activity, metal ion cofactors, 50
Catalytic (CAT) domain
  MMP, 38–40
  MMP-1, 45
Catechol oxidase(s)
  enzyme nomenclature into, 3–5
  oxo complex, 19
  PDB, published structures in, 22–27
  reactions catalyzed by, 4*f*
  vs. tyrosinases, structural differences, 29–30
Catechol oxidase from *Aspergillus oryzae* (AoCO4), 27
CD. *See* Circular dichroism (CD)
Cefotaxime hydrolysis, 55*f*
Chemical shift mapping method, 121–122
ChloroP 1.1 prediction, transit peptide, 9–10, 11*t*
Circular dichroism (CD)
  spectroscopy, 18
  and ZnMP, 129–131
Collagen, hydrolysis, 37–38
Collagenolysis mechanism, 43–46
Conformational diseases, 116–120
Core domain, PPOs
  conserved amino acid motifs, 8–9
  proteolytic cleavage site, 7–8
Coreopsis grandiflora, aurone formation, 3–5
CphA
  crystal structure of, 53
  MβLs, 51*f*

159

C

C-terminal domain, PPOs, 5
  conserved amino acid motifs in, 9
  proteolytic cleavage site, 7–8
Cu-binding domains, 8–9
Cysteine, 51–52

D

D256E structure, Pol β, 95–97
DNA polymerase β, 87–98
  application, 98–103
  broader picture, 97–98
  human prechemistry complex, 91f
  incorrect insertion, QM/MM study, 94–95
  molecular cluster model, 88f
  prechemistry active site, 92f
  QM/MM study of, 89–94
  X-ray crystal structure, 89, 95–97
DNA repair enzymes, 85
Double-Lewis activation, 54–56
Dpo4 NTP insertion, 100–101

E

Elizabethkingia meningoseptica, GOB-18 enzyme, 53
Enterobacter aerogenes, GpdQ, 50
Enzyme
  activity modulation, 127–128, 135
  with EDTA, 129–131
  folding, 124
  unfolding, 125–126
Escherichia coli
  agmatinase, 70–71
  MetAP, 58–59
  of polyphenol oxidase, 3–5

F

Fungal PPOs, 16–17
Fungal protyrosinase, 29
Fungal tyrosinase
  and cytoplasmic enzymes, 10
  from *Neurospora crassa*, 8
  from *Pycnoporus sanguineus*, 8
  sequence identities, 15

G

Glycerophosphodiesterase (GpdQ)
  active site structure, 61–62, 62f
  binuclear enzyme, 65
  bioremediation application, 63–65
  from *Enterobacter aerogenes*, 50, 60–61
  inactive mononuclear state, 62–63
  mechanism of, 64f
  α metal ion, 63
  metal ion composition of, 61–62
  organophosphate-degrading enzyme, 53–54, 60, 70
  role of, 60–61
GOB-18 enzyme, 53
Gold nanoparticles (AuNPs), ZnMPs, 128–129
GpdQ. See Glycerophosphodiesterase (GpdQ)

H

Hemocyanins, 3–5
  oxo complex, 22
  PDB, published structures in, 22, 24t
  X-ray crystallographic structural data, 22–30
Hemopexin-like (HPX) domain
  MMP, 38–40, 42f
  MMP-1, 45
Heterogeneity, in MMPs, 46–47
Histidine, 51–52
HPLC-MS, peptide detection, 132–135, 134f
HPX domain. See Hemopexin-like (HPX) domain
Hydrogen/deuterium exchange (HDX), 131–132
Hydrogen-deuterium exchange mass spectrometry (HDX-MS), 41
Hydrolyzed collagen, 37–38
HYP2 path, 103–105

I

Insulin-degrading enzyme (IDE) activity, 129–131
  chromatographic areas, 134f
  conformational changes in, 130f

M

Magnetic circular dichroism (MCD) spectroscopy, 62–63
Mammalian agmatinase, 71

Subject Index

Mass spectrometry (MS)
  advantages, 132–135
  biomolecules investigation by, 131–132
  and ZnMPs, 120, 131–135
Matrix metalloproteinases (MMPs), 116–120
  in cancer progression, 127–128
  and collagen hydrolysis, 37–38
  full-length, collagenolytic, 38–41
  heterogeneity in, 46–47
  ribbon representation, 39f
  structural evaluation of, 41–43
Maximum occurrence (MO), MMP-1 and, 40–41
MβLs. See Metallo-β-lactamases (MβLs)
MCD spectroscopy. See Magnetic circular dichroism (MCD) spectroscopy
Melanins, 6
Metal-binding site, 121–122
Metal ion cofactors, for catalytic activity, 50
Metallo-β-lactamases (MβLs)
  and antibiotics, 56–57
  for bacterial infections, 50–51
  binuclear, 54–56
  B1-type, 51–52, 56–57
  B2-type, 53, 56–57
  B3-type, 53
  B4-type, 53–54
  metal ion composition of, 51–52
  mononuclear reaction mechanism, 55f
  Zn1 site of, 51–52
Metalloproteins, 2–3
Methionine aminopeptidase (MetAP)
  binuclear metal center, 60
  binuclear reaction mechanism, 59f
  crystal structures of, 58
  Mn(II)-containing, 59–60
  mononuclear reaction mechanism, 58–59, 59f
  structural rearrangements, 58–59
  Zn(II) and, 57–58
Met tyrosinase, 17–18
MMP-1, 131–132
  CAT domain, 45
  HPX domain in, 45
  interaction with collagen, 43–44
  interaction with THP, 44–45
  and MO, 40–41, 45f

Phe301, 41–43
X-ray crystallographic structure, 42f, 47
MMP-3
  from C-terminal, 132–135
  His224 of, 124
MMP-8, 43
MMP-9
  SPR, 126
  on triple-helical collagen fragments, 45–46
MMP-12, 124–125, 127f
  heterogeneity, 46–47
  two-peptide intermediate, 44f
  X-ray crystallographic studies, 43
MMPs. See Matrix metalloproteinases (MMPs)
MO. See Maximum occurrence (MO)
Molecular cluster model, for Pol β, 88f
Mononuclear reaction mechanism
  for cefotaxime hydrolysis, 55f
  MetAP, 59f
MS. See Mass spectrometry (MS)
Mushroom tyrosinase, 8

N

Near-infrared (NIR) light, 127–128, 128f
Neurospora crassa
  fungal tyrosinase from, 8
  oxytyrosinase from, 19–22
NMR spectroscopy. See Nuclear magnetic resonance (NMR) spectroscopy
Nonblue copper centers, 2–3
N-terminal transit peptide, PPOs, 7
NTP insertion, Dpo4, 100–101
Nuclear magnetic resonance (NMR) spectroscopy
  and X-ray crystallography, 123–124
  ZnMPs, 121–123
Nucleic acid polymerases, 106–109
Nucleophilic hydroxide species, 54–56

O

ONIOM method
  in critical catalytic region, 86
  in Gaussian 03, 93, 104f
  of Morokuma group, 85–86
  QM/MM and, 91–92, 94–95
  quantum region size, 86

O3' proton
  and Asp256 position, 107–109
  hydrogen-bonded base oxygen, 107t
  original modeled position, 88
  to oxygen atoms, 102–103
  sugar, 87
  transfer, 91–92, 93, 94–95
Oryzae sativa, and PPOs, 9–10
Osteoporosis, PAP, 65–69
Oxo complex, 17–22
  aureusidin synthase, 22
  catechol oxidase, 19
  hemocyanin, 22
  investigation, 18
  PPOs, 20t
  tyrosinase, 19–22

P

PAPs. *See* Purple acid phosphatases (PAPs)
Paramagnetism-assisted NMR, 121–122
PDB. *See* Protein Data Bank (PDB)
Peptide
  and amyloid fibrils, 116–120
  detection by HPLC-MS, 134f
  fragment, 131–135
  ZnMP and, 129
Pholiota nameko, mushroom tyrosinase, 8
Phylogenetic tree
  of polyphenol oxidases, 12f
  uniprot for, 13t
Plant PPOs, 16
PO3, 88
Pol κ insertion, 101–103
Pol λ insertion, 98–100
Polynuclear metalloenzymes, 50
Polyphenol oxidases (PPOs), 3–5
  core domain, 7–9
  C-terminal domain, 5, 7–8, 9
  general sequence structure
  mutants, 16–17
  N-terminal transit peptide, 7
  oxo complex of, 20t
  phylogenetic tree of, 12f
  sequence homologies within, 11–15
  transit peptide and location, 9–10
  in vitro activation, 16
  in vivo activation, 16

X-ray crystallographic structural data, 22–30
Protein(s)
  met form, 3
  misfolding, 116–120
  oxy form, 3
Protein data bank (PDB)
  catechol oxidases, 22–27
  hemocyanins, 22
  published structures in, 22–29
  tyrosinases, 27–29
Purple acid phosphatases (PAPs)
  active site of, 66f
  binuclear metallohydrolase-catalyzed esterolysis, 67f
  comprehensive model, 67–68
  crystal structures of, 66–67
  enzyme regeneration, 68–69
  Fe(III) and, 70
  in mammals, 65–66
  and μ-hydroxide, 67–68
  overabundance, 69
  plant, 65–66
Pycnoporus sanguineus, fungal tyrosinase, 8
Pyrococcus furiosus, MetAP, 57–58

Q

Quantum dot (QD), for cancer detection, 128–129
Quantum mechanical (QM) calculation
  on Pol β prechemistry complex, 89
  at proposed reaction path, 87–88
Quantum mechanical/molecular mechanical (QM/MM) study
  application, 98–103
  describing reactive pathways, 85–87
  nucleic acid polymerases, 106–109
  of Pol β, 89–94
  Pol β incorrect insertion, 94–95
  Pol κ insertion, 101–103
  Pol λ insertion, 98–100
  X-ray crystal structure of Pol β, 95–97

R

Resonance Raman spectroscopy, 18
RNA polymerase, 103–106

S

Site-directed mutagenesis
  in bacterial PPOs, 17
  on fungal tyrosinase, 16–17
  in plant PPOs, 16
Small-angle X-ray scattering (SAXS) data, 40
Sodium dodecyl sulfate (SDS), PPOs, 16
Stenotrophomonas maltophilia, MβLs, 51*f*
Streptomyces castaneoglobisporus, bacterial tyrosinase, 27
Surface plasmon resonance (SPR), ZnMPs, 125–126

T

TargetP 1.1 prediction, transit peptide, 9–10, 10*t*
THP models. *See* Triple-helical peptide (THP) models
Thylakoid transfer domain (TTD), 9
Transit peptide
  ChloroP 1.1 prediction, 9–10, 11*t*
  PPOs, 9–10
  TargetP 1.1 prediction, 9–10, 10*t*
Triple-helical peptide (THP) models, 41–43, 42*f*, 44–45, 47
Triple-resonance 3D NMR experiments, 122
Type-3 copper, 3–5. *See also* Polyphenol oxidases (PPOs)
  active site, 5*f*
  aureusidin synthase role, 6
  electronic properties of, 17–18
  hemocyanins, 22
  UV absorption, 18
  X-ray structures of, 30
Tyrosinase(s), 3–5, 4*f*
  vs. catechol oxidases, structural differences, 29–30
  fungal, 8
  in melanins, 6
  mushroom, 8
  oxo complex, 19–22
  PDB, published structures in, 27–29
  structural data list, 24*t*
  X-ray crystallographic structures, 5–6
Tyrosinase from *Bacillus megaterium* (BmTYR), 28

U

Uniprot
  for phylogenetic tree, 13*t*
  and PPOs, 7
UV/vis absorption spectroscopy, 18

V

Vitis vinifera, CO structure, 23–27

X

X-ray crystallography
  MMP-12 and MMP-8, 85
  PPOs and hemocyanins, 22–30
  ZnMPs, 123–125
X-ray crystal structure
  active site of, 93, 96*f*
  calcium ions and, 100–101
  equilibration, 94–95
  examination, 94–95
  hydrogen-bonded base oxygen in, 107*t*
  Pol λ insertion, 98–100, 99*f*
  QM/MM study based on, 95–97

Z

Zinc metalloproteases (ZnMPs)
  co-crystallization of, 124
  and conformational diseases, 116–120
  kinetic parameters, 125–126
  mass spectrometry and, 131–135
  monitoring, 135
  NMR spectroscopy, 121–123
  optical sensing, 127–128
  substrate, 120
  surface plasmon resonance, 125–126
  X-ray crystallography, 123–125
Zn-binding amino acidic residues, 123
Zn(II)-bound hydroxide, 54

Printed by Printforce, the Netherlands